ENERGY USE WORLDWIDE

Selected Titles in ABC-CLIO's
CONTEMPORARY
WORLD ISSUES
Series

For a complete list of titles in this series, please visit
www.abc-clio.com.

Books in the Contemporary World Issues series address vital issues in today's society, such as domestic politics, human rights, and homeland security. Written by professional writers, scholars, and nonacademic experts, these books are authoritative, clearly written, up-to-date, and objective. They provide a good starting point for research by high school and college students, scholars, and general readers as well as by legislators, businesspeople, activists, and others.

Each book, carefully organized and easy to use, contains an overview of the subject, a detailed chronology, biographical sketches, facts and data and/or documents and other primary-source material, a directory of organizations and agencies, annotated lists of print and nonprint resources, and an index.

Readers of books in the Contemporary World Issues series will find the information they need in order to have a better understanding of the social, political, environmental, and economic issues facing the world today.

ENERGY USE WORLDWIDE

A Reference Handbook

Jaina L. Moan and
Zachary A. Smith

**CONTEMPORARY
WORLD ISSUES**

A B C ☙ C L I O

Santa Barbara, California
Denver, Coloirado
Oxford, England

Library of Congress Cataloging-in-Publication Data
Moan, Jaina L.
  Energy use worldwide : a reference handbook / Jaina l. Moan and
Zachary A. Smith.
  p. cm. — (Contemporary world issues)
  Includes bibliographical references and index.
  ISBN 978-1-85109-890-3 (hard copy : alk. paper) —
  ISBN 978-1-85109-891-0 (ebook) 1. Power resources—Handbooks,
manuals, etc. 2. Energy consumption—Handbooks, manuals, etc. I.
Smith, Zachary A. II. Title.
  TJ163.2.M62 2007
  333.7913—dc22
2007007414
11 10 09 08 07   1 2 3 4 5 6 7 8 9 10

ABC-CLIO, Inc.
130 Cremona Drive, P.O. Box 1911
Santa Barbara, California 93116-1911

This book is also available on the World Wide Web as an eBook.
Visit abc-clio.com for details.

This book is printed on acid-free paper. ∞

Manufactured in the United States of America

This book is dedicated to Benjamin Moan.
Thank you for all of your love and support.

Contents

List of Figures

List of Tables

1

Background and History

Introduction

Energy is an essential part of our world. Plants depend on solar energy to grow; our bodies depend on food energy to maintain their metabolism; our society depends on energy for electricity, transportation, and industry. This chapter provides an overview of the fundamental aspects of energy: what it is, where it comes from, how it is measured, why it is important to society, and the historical development of energy resources globally. The first part of this chapter describes the physical properties and fundamental concepts of energy. The second part of the chapter discusses renewable and nonrenewable sources of energy and how these sources are converted into energy used by society. Finally, a third part highlights important historical events in energy use.

Energy Concepts

Because energy makes up such a large part of our world, it is important to understand the basic physical concepts of energy and where it comes from. This section examines physical definitions, energy conversion and efficiency, electricity generation, and energy units. These topics are fundamental in the disciplines of physics and engineering. Physics is a subject that explains many of the energy dynamics observed in our world. Engineering is a field that utilizes physical laws to design systems for harnessing and distributing energy to society. These concepts are important for understanding how energy resources are used and consumed in our society.

Physical Definitions

The meaning of energy embodies many concepts and means different things to different people. Because of this complexity, it is impossible to give a set definition for energy. However, the generally agreed upon physical description of energy is "the capacity to do work" (Smil 1999, xiii). In order to understand what this means, the concepts of force and work must be described.

Mathematically, force is the product of an object's mass and its acceleration.

Force = mass x acceleration (change in velocity over time)

Essentially, force is the phenomenon that causes an object to change its motion (Wolfson and Pasachoff 1995, 95). Work, then, is defined as the product of force and distance.

Work = Force x Distance

In other words, in order to quantify mechanical work, one must first measure the amount of force that was applied to a given object and multiply it by the distance that the object moved. The number given for this measurement is equivalent to the amount of energy used to move the object and the value is expressed in joules (J).

Work and force are simple equations useful for understanding that energy is observable and can be measured by the forces exerted on an object in motion. There are two forms of energy. Kinetic energy is energy that is moving. Electrical and thermal energies are examples of kinetic energy. Another form, potential, is the energy that is stored in objects. Chemical (the energy stored in chemical bonds) and stored mechanical energy (e.g., the energy stored in water held by a dam) are two examples of potential energy. Distinguishing between these two forms of energy is important because society extracts useful work by converting energy from one form (potential) to another (kinetic). For example, when coal is burned, or combusted, its chemical energy is released in the form of heat (steam). The steam turns large turbines to produce mechanical energy, which is then converted by a generator to electrical energy. Similarly, when stored water from a dam is released, the falling water turns large turbines producing mechanical energy. Efficient energy conversion is fundamental to society's ability to harness energy from primary sources. The next section examines the energy laws associated with this process.

Energy Conversion and Efficiency

Energy conversions are processes that determine how energy is harnessed from sources like coal or solar radiation to serve the needs of society. When energy is converted from one form to another, it is constrained by physical laws, or the laws of thermodynamics. The first law is the conservation of energy. This law states that energy cannot be created or destroyed; it can only be converted from one form to another. In society, consumption is a term that is used to describe the process of conversion. Energy is not actually created or destroyed in the process of consumption; it is converted from one form to another (Ramage 1997, 98).

The second law states that although energy is never destroyed, it does decrease in quality. As energy is converted from one form to another, the amount of useable energy in the system declines and more energy is needed to extract the same amount of mechanical work. In every energy system (one that utilizes energy conversions from its initial state to its final end use), all energy ends up as waste heat. This process is not reversible. That is, the useful energy obtained can never be captured and reused as it was in its stored form. Hence, the second law of thermodynamics states that as a system converts energy to a useful form, the system becomes more entropic, or disorganized, and the resulting energy is less useful for doing work.

Another important aspect to the second law of thermodynamics is that as a system converts energy from one form to another, it is not possible to extract the same amount of energy in the form of work that is contained in the system (Wolfson and Pasachoff 1995, 528). In any system, some energy will inevitably be lost as heat energy. The system can never be 100 percent efficient. Because of this, the energy efficiency, or the ratio of useful energy output to total energy input, is a valuable measure for understanding how much energy can be harnessed from a particular source.

Energy efficiency is an important value to quantify because different conversion processes have different efficiencies. The most efficient systems are those that can directly convert potential (or stored energy) into useable energy without the input of additional energy, such as heat. For example, the motion of falling water is a much more efficient energy conversion than the burning of coal. Water only needs to fall from a high point to a low point to release energy. Coal, on the other hand, needs to be heated in the presence of oxygen (or combusted) in order to release its chemical energy. This process not only requires the addition of heat energy to combust the fuel, it also releases a large amount of energy as waste heat. Any energy system that relies on the addition of heat energy is much less efficient in converting its input into heat energy.

Electricity

Electricity is a very important secondary energy source. It is generated from primary sources (e.g., fossil fuels) and is used for many purposes; electric appliances, lighting, heating, and cooling all are powered by electricity. The physical properties of electrical energy allow for its transmission across long distances from its source of generation. This section discusses the fundamental aspects of electrical energy, magnetism, and transformers. These concepts describe how electricity is generated and transported.

Electrical energy is primarily derived from electrons, very small particles that orbit around the nuclei of atoms and are held to the nucleus with an electric force. Certain elements, like metals, have a large amount of electrons that orbit their nuclei. The electrical energy that holds these particles to the nucleus can be released with the introduction of a charge. When this happens, electrons become disassociated from the atoms and move freely within the matrix of the element. Metals, like copper, are good conductors of electricity because they contain large amounts of electrons that become dissociated easily from their atoms with the application of an electrical force (Ramage 1997, 153). When this force travels along the length of a wire, it is called a current. When the ends of the wire are connected in a closed path, the current creates a circuit and electrical energy can be used to light homes and power appliances.

The concept of magnetism is also important for the generation of electricity. Magnetism is a property found in iron, or materials that attract iron, that exerts an attractive or repulsive force on other objects with magnetic fields (Wolfson and Pasachoff 1995, 723–724). It is thought that magnetic forces are generated from the quantum mechanics that define the structure of atoms and nuclei. Magnetism is important because it interacts with electrical forces to produce an electric current. A generator, which is a machine that produces electrical energy from mechanical energy, produces an electromagnetic current by passing a coil of conductive wire past the positive and negative poles of a magnet.

The concept of induction describes how electricity is transported from its source to its final end use. Induction is the process by which electrical current can be generated in a charged circuit from an adjacent charged circuit by proximity and grounding (Wolfson and Pasachoff 1995, 852). Transformers are devices that embody the concept of induction and allow for electricity to travel long distances. A transformer consists of two or more coils of wire that are situated in such a way that a secondary wire can pick up the charge of a primary wire carrying electric current. The transformer can also increase or decrease the voltage that is flowing through a wire. This feature of transformers is useful for distributing safe amounts of electricity from high-voltage wires.

The fundamental ideas behind electrical energy and magnetism can be applied to illustrate how an electrical power plant generates electricity. Electricity is made from primary sources of energy, such as coal combustion or wind power. For example, a coal-fired power plant combusts coal to create hot steam. The hot steam turns large turbines that are connected via a long shaft to a generator. The generator contains a magnet. The turning shaft from the turbines has a long metal coil wrapped around it. As the coil turns between the positive and negative poles of the magnet, an electrical current is generated. This current is transmitted along high-voltage power lines to substations that contain transformers. The substations then release low-voltage electricity to distribution lines in communities where it is used. When a light switch is turned on, a circuit is connected to the electrical power in the wires and light is provided.

Energy Measurement and Units

Because energy is such an important part of our lives, it is important to understand the value of energy units. Units are a way of measuring and quantifying how much energy is available, produced, and consumed in our society. This section provides a working understanding of what energy units are and how to interpret them.

Energy values can either be expressed in basic physical units (e.g., joules), or in units that refer to a particular energy source (e.g., barrels of oil equivalent). The magnitude of units is often portrayed in metric scale, so it is important to grasp how different values are described when they increase or decrease in size. For example, 1 million joules is equal to 1 megajoule (MJ), and 1 billion joules (J) is equal to 1 gigajoule (GJ). Table 1.1 describes basic metric conversion factors between magnitudes of units.

The joule is the standard unit of energy according to the International Standard (SI) system of units. One joule is a physical unit of energy that describes how much work is done on a system when an applied force of one newton is required to move an

TABLE 1.1
Metric Conversion Factors

Prefix	Abbreviation	Scientific notation	Name	Value
Deci	D	10^{-1}	Tenth	0.1
Centi	C	10^{-2}	Hundredth	0.01
Milli	M	10^{-3}	Thousandth	0.001
Micro	μ	10^{-6}	Millionth	0.000001
Nano	N	10^{-9}	Billionth	0.000000001
Pico	P	10^{-12}	Trillionth	0.000000000001
Femto	F	10^{-15}	Quadrillionth	0.000000000000001
Atto	A	10^{-18}	Quintillionth	0.000000000000000001
Deka	Da	10^{1}	Ten	10
Hector	H	10^{2}	Hundred	100
Kilo	K	10^{3}	Thousand	1,000
Mega	M	10^{6}	Million	1,000,000
Giga	G	10^{9}	Billion	1,000,000,000
Tera	T	10^{12}	Trillion	1,000,000,000,000
Peta	P	10^{15}	Quadrillion	1,000,000,000,000,000
Exa	E	10^{18}	Quintillion	1,000,000,000,000,000,000

object one meter. (A newton is the standard unit of force.) The joule also describes how much energy is stored in a particular object. For example, the amount of energy stored in a barrel of crude oil is approximately 6 GJ. In other words, 6 billion joules of energy can potentially be extracted from a barrel of oil (Smil 1999, xiv). However, because of the second law of thermodynamics, it would be impossible to convert 100 percent of the potential energy into useable energy.

Another unit used to describe energy quantities is the British thermal unit (Btu). This unit is often used to express the heat energy content of fuels (e.g., coal), and it is defined as "the quantity of heat needed to raise the temperature of one pound of water by one degree Fahrenheit" (EIA 2003). The definition of a Btu is better understood as being a measure of energy stored in an object. Used the same way a joule is, one Btu is equivalent to 1,055 joules. So, one barrel of oil (which contains 6 GJ of energy) contains approximately 5,687,204 Btu of energy.

Other units that are used to describe amounts of energy are the calorie and the kilocalorie (kcal, which is 1,000 calories). The calorie is defined as the amount of energy required to heat one gram of water one degree Celsius. The calorie is a measure of energy used to describe the energy released in chemical reactions (Wolfson and Pasachoff 1995, 165). This unit is also used for determining the amount of energy that is contained in food. An adult human male, for example, needs to consume approximately 2,500 kcal per day. Since 1 kcal is equal to 4,200 joules, this energy requirement is approximately 10 MJ, or 10 million joules of energy (Smil 1999, xv).

Rates are a way of expressing how much energy society is consuming in a given amount of time. The rate at which energy is converted to useable forms of energy is called power. The watt, which equals one joule per second, is the unit that describes this rate. So, a 500-watt generator converts mechanical energy to electrical energy at a rate of 500 joules per second. A large coal-fired power plant generates electricity (converts mechanical energy to electrical energy) at a rate of 500–700 megawatts (MW, or 1 million watts), or 500 million joules per second (Ramage 1997, 161). The kilowatt-hour is another common unit for energy rates. It describes how many kilowatts of electricity are used in one hour. The KWh is the typical unit of measurement that power companies use when billing for electricity.

Energy units are also expressed in terms of the type of fuel they quantify. The petroleum industry measures energy by tonnes of oil equivalent (toe) or barrels of oil equivalent (boe). A standard barrel of oil contains 42 U.S. gallons, or 159 liters. There are approximately 7.3 barrels of oil in a tonne, so approximately 41.9 GJ of energy are contained in one tonne. Tonnes of coal equivalent (tce) is a measure that is used to describe the energy in coal. The amount of energy in a tce can vary because of different coal types, but the value of 29 GJ per tonne is accepted as an international standard (Ramage 1997, 13). Table 1.2 describes unit conversions of different energy units in terms of joules. Energy units also describe quantities of energy resources. Oil is measured in barrels of crude. Coal is measured in tonnes, or short tons (one short ton equals 2,000 pounds, or 907.2 kilograms). Natural gas is measured in cubic feet. Society often describes resource availability and consumption quantities using these units.

Sources of Energy

Humans use a vast amount of energy. In 2002, the world consumed 412 quadrillion Btus of energy, which is equivalent to approximately 435 EJ (EIA 2004b, 298). Most of the primary energy sources used today are nonrenewable. Approximately 85 percent of all energy produced and consumed is derived from finite supplies of fossil-fuel primary-energy sources. The remaining 15 percent of energy comes from nuclear and renewable sources (294).

TABLE 1.2
Energy Equivalents

Unit	Equivalent amount
1 Btu	1055 J
1 calorie	4.2 J
1 kcal	4200 J or 4.2 kJ
1 kilowatt-hour	3,600,000 J or 3.6 MJ
1 boe	6,000,000,000 J or 6 GJ
1 toe	41,868,000,000 or 41.9 GJ
1 tce	29,000,000,000 J or 29 GJ.
1 watt	1 joule/second

Nonrenewable energy sources are those that become depleted with use and cannot be replenished within a reasonable amount of time. A renewable energy resource is defined as natural energy flows that are not depleted with use and can be regenerated as they are depleted (Alexander 1996, 27). It is important to note the difficulty in measuring exact values for the production and consumption of energy from different primary sources. Commercially traded sources provide the best data since they have a market value and hence quantity is tracked. Other sources, such as biomass, are more difficult to measure because they are not traded on a commercial basis.

This section discusses the characteristics of primary energy sources: what they are, where they are found, and how energy is harnessed from each resource. Fossil fuels and nuclear sources (the nonrenewable sources) are discussed first since they provide such a large portion of energy needs. Then, because of its future importance, renewable energy is examined.

How Does Society Use Energy?

Before describing the various ways in which energy can be harnessed, it is important to understand how energy resources are used in society. There are four primary end uses of energy: industrial, residential, commercial, and transportation applications. In the industrial sector, energy is used to make metal and paper, for petroleum refining, agriculture, the chemical industry, and the manufacturing industry. This sector comprises approximately 33 percent of the energy used in a developed society. The residential sector uses energy in homes for heating and cooling, lighting, electrical appliances, and water heating. This sector comprises 22 percent of the energy used by society. The commercial sector uses energy for much of the same applications as the residential sector. Heating, cooling, and lighting are the main uses of energy in restaurants, retail and office buildings, schools, hospitals, and churches. Commercial energy uses comprise 18 percent of energy consumed in society. Finally, transportation is the fourth sector. All vehicles use some form of energy to move from one place to another, and most of this energy is derived from fossil fuels. This sector comprises 27 percent of energy used by society (EIA 2004b).

It is important to note that the energy distribution to each sector is different in every country. The percentages listed above correspond to the United States. Similar patterns exist in other developed countries. In general, developed countries allocate more energy to industrial and transportation sectors. Less developed nations allocate more of their energy consumption to domestic uses. Additionally, the primary sources used to meet the energy needs of industrialized nations are different from those consumed in developing countries. These energy dynamics and their implications are discussed further in chapter 2.

Fossil Fuels

Fossil energies are extracted from beds of once-living organic matter (primarily plant) that was compressed among and between layers of rock throughout geologic history. The heat and pressure caused by compression in different types of rock layers formed the different types of fossil fuels. The composition of these fuels is primarily made up of carbon, oxygen, and hydrogen, but depending on the fossil fuel type, may contain many other elements and impurities. Hydrocarbons, which are molecules composed of carbon and hydrogen atoms, are a group of important compounds associated with these fuels. Fossil fuels provide heat energy when they are burned (or combusted). The resulting heat is converted to mechanical energy by the use of combustion engines (as in the case of vehicles) or to electricity by turbines and electric generators (as in the case of power plants).

This section examines the general characteristics of coal, oil, and natural gas. It describes the extraction, processing, and transportation of each of these fuels, followed by a brief overview of estimated global totals of reserves (the amount of a particular resource that is estimated and recoverable) and consumption. Energy statistics are presented in greater detail in chapter 6.

Natural Gas

Natural gas is 80 to 95 percent methane (CH_4), which is a simple fuel containing one carbon atom and four hydrogen atoms (Stoker, Seager, and Capener 1975, 113). In its natural state in the environment, natural gas deposits may also contain heavier hydrocarbon impurities (such as propane or butane), water, carbon dioxide, and hydrogen sulfide. Seismic and drilling explorations are used to reveal the potential sites that contain natural gas.

Once these sites are discovered, natural gas is extracted from the subsurface by drilling a well. Gas is piped to a processing plant where hydrocarbon impurities are removed with heavy oils, water is removed with drying agents, hydrogen sulfide compounds and carbon dioxide are removed, and finally an odor agent is added to the processed gas for purposes of leak detection. Natural gas is generally transported by pipeline from processing plants to areas of use. An extensive natural gas pipeline network lies across large land areas. An alternative method of storage and transportation is made possible by compressing natural gas into liquefied natural gas (LNG), which reduces the volume of the gas by 600 times. LNG operations cool the gas to a liquid (–259 degrees Fahrenheit; –162 degrees Celsius), and then re-gasify it when it reaches its destination or when demand for natural gas is higher (135).

Natural gas is the least consumed of all the fossil fuels, accounting for approximately 23 percent of energy production in 2002 (EIA 2004b, 300). It is estimated that global recoverable reserves total anywhere between 6,040 and 6,805 trillion cubic feet (EIA 2005). In 2004, approximately 91.76 trillion cubic feet of natural gas was consumed in the United States, which comprised approximately 25 percent of global natural gas consumption (EIA 2004b, 316). It is estimated that the use of natural gas will increase in the future as prices of petroleum rise and the undesirable effects of coal reduce that source's demand (Smil 2003, 213).

Petroleum (Oil)

Petroleum is composed of a complex mixture of hundreds of different hydrocarbons. Petroleum may also contain impurities, such as sulphur, nitrogen, oxygen, and trace amounts of metals. Because of the complexity of its composition, refining is necessary for getting it into a useable form. There are many useable products that petroleum resources provide. Gasoline, jet fuel, kerosene, and lubricants are a few of the commercial substances extracted from petroleum.

Crude oil, a thick, viscous fluid, is extracted from the ground by drilling and pumping. It is then transported either by ship or pipeline to a refinery where the different components of the crude oil are partitioned using a process called distillation, which separates out the hydrocarbon compounds using their different boiling points. Secondary conversion processes, such as thermal and catalytic cracking, chemically transform less useful fractions of hydrocarbons into marketable commodities (e.g., gasoline). These processes break large hydrocarbon molecules into smaller constituents. Petroleum is then purified to remove any impurities that produce harmful substances when burned. Gasoline and jet fuel are the most marketable products of petroleum, and they are used mainly for transportation purposes.

Global production of petroleum has risen drastically since 1950. It is now the most utilized energy resource, comprising 37.7 percent of global energy production in 2003 (153 quadrillion Btus in 2002) (EIA 2004b, 294). The United States is by far the largest user of petroleum, consuming 19.8 million barrels of oil per day (312). This consumption is supported from both domestic and foreign sources. The largest reserves of petroleum, estimated at 670 to 690 billion barrels, are found in the Middle East (300). Social tensions arising from resource availability, limited supplies of recoverable petroleum, and environmental effects of fossil fuel combustion may limit the use of this resource in the future.

Coal

Coal is the most chemically complex fossil fuel that is burned for energy purposes. Although it consists mainly of carbon, the chemical structures within coal matrices contain significant concentrations of nitrogen and sulfur, and trace amounts of many other elements, including mercury, lead, and other metals that are toxic to humans. Volatile gases and water are also bound within coal's chemical structure, and their release to the atmosphere during combustion can be very harmful to human health and the environment. Coal was formed from the fossilization and compression of large swampy areas or peat bogs. Different coal types were formed from varying degrees of heat and pressure exerted on the organic matter in these environments over long periods of geologic history. Coal types are ranked according to the amount of fixed carbon and volatile matter; the higher the rank of coal, the greater the amount of fixed carbon and the lower the amount of volatile matter (Miller and Miller 1993, 28). Figure 1.1 describes the rank of coal from lignite to anthracite.

Coal has many uses in society. Anthracite is a high-ranking coal that is used mostly for domestic heating purposes. Bituminous coal is primarily used in electricity generation and coke

FIGURE 1.1
Different Ranks of Coal

Rank			
Low			**High**
Lignite – "Brown Coal"	Bituminous coal	Semianthracite	Anthracite – highest
C: 30 to 55 percent	C: 48 to 73 percent	C: 83 to 90 percent	ranking. Similar to
VM: 18 to 20 percent	VM: 30 to 40 percent	VM: 10 to 15 percent	semianthracite, but
M: 30 to 43 percent	M: 3 to 11 percent	M: NA	less friable

C = fixed carbon, VM = volatile matter, M = moisture

Source: Miller and Miller, 1993, 28

production. (Coke is produced from the pyrolysis of coal. Because it has a higher heat value than coal, it is used as a fuel source for iron ore smelting for steel production.) Coal is extracted from the ground by a variety of different mining techniques. Deep shaft mining is used in areas where coal seams are located 100 feet or greater below the surface. In other regions, where coal is located closer to the surface, the land is stripped away to reach the coal beds below. This is called strip or surface mining, and while it is safer for miners, it is devastating to the landscape.

After extraction, coal bound for power plants is pulverized before being transported. The power plant blows the coal dust into a furnace in the presence of oxygen. The hot gas that is created from the combustion process is directed into a boiler containing water pipes. The water is heated from the hot gas to create steam, which is directed to a turbine electric generator. Steam leaving the generator is cooled and condensed back into water and transported back to the boiler (Stoker, Seager, and Capener 1975, 161). Electricity generated from coal combustion is transported via high-voltage power lines to areas where it is needed.

The largest global coal reserves are found in the United States (272 billion short tons), Russia (173 billion short tons), and China (126 billion short tons) (EIA 2004b, 318). Global consumption of coal in 2002 was 5,262 million short tons (EIA 2004c). China was the largest consumer, using approximately 27 percent of the global total (EIA 2004b, 322). Despite the large amounts of coal reserves, global demand for coal has declined in the past fifty years because of its undesirable environmental effects and the availability of more concentrated stores of energy found in nuclear fuel.

Nuclear

The conversion of energy from nuclear primary sources also creates steam to power an electric generator, but the main difference occurs in how the energy is released from the fuel. With fossil fuels, the process of combustion releases chemical energy that is stored in the chemical bonds between molecules in the fuel. A nuclear reaction, on the other hand, releases energy contained in the nuclei of atoms.

To understand a nuclear reaction, it is necessary to define the structural parts of an atom. Atoms are the smallest components of any given element. They are made up of a nucleus that contains protons, neutrons, and a system of electrons that exists outside the nucleus. Protons and neutrons together make up most of the mass of an atom. (An element's atomic number is calculated by summing numbers of protons and neutrons in the nucleus, while its atomic weight is calculated from the mass of the protons, neutrons, and electrons.) In general, most atoms of a particular element have the same number of protons and neutrons, but many elements have isotopes, which are atoms of the same element that contain more neutrons than protons in their nuclei. Isotopes can either be stable (do not release energy, or decay, over time) or radioactive, meaning their nucleic structure is unstable and decays over time releasing energy. This spontaneous nuclear reaction is the process by which new elements are formed.

The nuclear reaction can be manipulated in order to produce forms of energy that are useful (or harmful) to humans. In order to harness this energy, neutrons are used to split radioactive atoms, a process called fission. The splitting of one atom releases additional neutrons that split additional atoms. Hence, a nuclear reaction is a sustained chain reaction that releases energy from atomic nuclei. Instead of using energy from chemical bonds (the process that occurs in fossil fuel combustion), a nuclear reaction utilizes energy contained in the nuclei of atoms.

Uranium is one fuel that is required in a nuclear reaction. Thorium and plutonium also are elements that sustain nuclear reactions, but this brief discussion focuses on the uranium fuel cycle since it is the main fuel used in nuclear reactors. Two uranium isotopes are important in the nuclear fuel cycle, U-235 and U-338. Uranium has 92 protons, so U-235 contains 92 protons and 143 neutrons, and U-238 contains 92 protons and 146 neutrons. Although U-235 is the same element, it exhibits vastly different characteristics. One of its characteristics is the ability to fission upon impact with other neutrons.

Uranium fuel is produced from uranium ore, whose largest quantities are found in the deserts of the southwestern United States. After the ore is mined, it is milled to produce U_3O_8, or "yellowcake." The yellowcake is then converted to its gaseous phase, uranium hexafluoride (UF_6), in preparation for enrichment. At this point, the uranium resource contains only about 0.71 percent U-235 and approximately 99.3 percent U-238 (Rose 1986, 287). In order to be effective in a nuclear reaction, it must be enriched so that it contains at least 3 percent U-235. Essentially, the process of enrichment works to increase the ratio of U-235 to U-238. The enriched fuel is then converted to uranium oxide (UO_2) in the form of small, ceramic pellets that are packed in zircaloy fuel rods. Zircaloy is a metal alloy consisting of zirconium, tin, chromium, and nickel known for its heat-resistant properties. The fuel rods are bundled into fuel assemblies and are used in nuclear reactors for electricity generation.

A nuclear reactor is composed of four parts: (1) the fuel rods described above; (2) control rods that control the rate of the reaction; (3) the coolant that carries the heat away from the reactor; and (4) the moderator that slows the speed of the reaction. Reactors normally contain between 100 and 300 fuel assemblies, which can operate continuously for approximately two years (EIA 2004a). "Spent" fuel rods are transported to a secured area for storage. Because of their high radioactivity, fuel rod assemblies are first stored in shallow pools of water so that short-lived, intense radioactivity can be reduced. The fuel rods are either then reprocessed to try to recover useable uranium or are moved to long-term storage.

Radioactive waste disposal and storage is difficult because high-level radioactive material is very harmful to human health (see chapter 2). Radioactive waste is stored with nitric acid solution in stainless steel tanks in many different locations. There have been efforts to find one single repository for all of the nuclear waste produced in the United States. Yucca Mountain in south-central Nevada was chosen as this site; however, the storage of nuclear material there has been delayed for many reasons, including issues of transportation and scientific integrity in site selection. Yucca Mountain is discussed in chapter 3.

Some 6.7 percent of the energy produced worldwide is from nuclear power. The United States is the largest producer of nuclear energy, generating approximately 780 billion kilowatt-hours in 2002 (EIA 2004b, 328), providing 20 percent of U.S. electrical energy needs. Western European countries also produce significant amounts of nuclear power. France obtains 78 percent of its electricity needs from nuclear energy (EIA 2004a). Belgium receives 55 percent of its electricity and Sweden harnesses 51 percent of its power from nuclear sources (NEA 2005). Japan also uses nuclear power for 30 percent of its electricity. Nuclear energy is likely to be considered more in the future as the concern over global warming increases. However, because of the negative effects of radioactivity and the lack of public acceptance for nuclear power, it remains to be seen what role nuclear power will play.

Renewable Sources

Renewable sources of energy are becoming increasingly important as potential energy resources. This section discusses five categories of renewable energies: solar (active, passive, and photovoltaic); water (hydroelectricity, tidal, and wave); wind; biomass; and geothermal. Globally, these resources comprise somewhere between 8 and 16 percent of primary energy use (EIA 2004a; Ramage 1997, 20). Most of the renewable energy used is in the form of hydroelectricity and biomass, with the remaining renewable sources contributing less than 1 percent.

Solar Energy

Most of the energy sources on the planet are indirectly derived from the Sun. It is estimated that approximately 170,000 terawatts (TW) of solar radiation is constantly impacting the surface of the Earth (Rose 1986, 71). Two-thirds of this radiation is reflected back into space, but the remaining energy is greater than one hundred times the amount of power presently available on Earth (Ingersoll 1990, 207). Although not all of this energy can be harnessed, the

Sun represents a potentially large primary source. Solar energy is the cause of many natural processes on Earth that provide renewable resources. The Sun provides the energy for photosynthesis to occur, resulting in the large amount of biomass resources. Solar radiation causes shifts in wind patterns and the hydrologic cycle, creating the potential for wind and water energy. Other sections of this chapter are devoted to those energy sources. This section focuses on active and passive solar technologies, solar thermal engines, and photovoltaics as ways of harnessing solar energy to meet the needs of society.

Solar thermal energy can be captured in either active or passive ways. Active solar heating uses a device called a solar collector to gather and concentrate solar radiation (Everett 1996, 41). Generally, active solar technologies are used for water and space heating applications. Passive solar technologies are also used for heating, but they incorporate building design elements that capture heat and light from solar radiation. In recent years, many advances have been made in passive solar designs that decrease a building's reliance on fossil-fuel-derived energy sources.

Unlike active and passive technologies for capturing solar radiation, solar thermal engines are a way to convert solar radiation into mechanical work for the production of electricity. This process uses mirrors to concentrate solar radiation for boiling water to create steam for electric generators. The first and largest thermal engine power plant was built in the Mojave Desert in California in 1984. It was operated by Luz International. The company went bankrupt and the plant closed in the 1990s, but during its operation, the plant had an electricity-generating capacity of 80 MW (Everett 1996, 78).

Photovoltaics (PVs) are another way of capturing solar energy. Photovoltaic cells convert sunlight directly to electricity using solid-state, crystalline materials (Boyle 1996, 92). Some materials, like selenium, exhibit electric properties when exposed to light. When these materials are crystallized with semiconducting elements (nonmetallic materials that are able to conduct electricity), like silicon, a PV cell is formed and electricity can be conducted. PV systems have the potential to supply power away from utility grids if needed and have been used to supplement power grids. For example, the German electric utility company RWE has used a PV plant to supply approximately 250,000 kilowatt-hours per year to its electricity grid (122).

Water Energy

The energy that is stored in water can be converted into electricity. Conversion is done using hydroelectric dams, capturing wave energy, and also by exploiting the tidal forces on the planet.

The energy that is provided by hydroelectric dams is indirectly supported from solar energy. Solar radiation hitting the Earth is the main driver of the hydrologic cycle, which is the geochemical cycle that recycles water among the land, water bodies, and atmosphere. Solar radiation drives the weather patterns that allow for rainfall and runoff to occur, making it possible to capture running water and harness its energy. Hydroelectricity provides 20 percent of the world's power, making it the most widely exploited renewable source of energy (Ramage 1996, page 181). A hydroelectric dam captures energy through large water turbines placed at the bottom of the dam to intersect the water as it falls from a high point to its low point. The turbines are connected to large electricity generators. The efficiency of this process is very high since it does not involve a heat engine.

Tidal power uses tidal forces—those that result from the moon's gravitational pull on the seas—as its driving force to move water. In order to exploit this force, large barrages, which are a type of dam, are constructed in estuaries for the purpose of capturing water as the tide rises. As the tide comes in, water flows through sluice gates. At high tide, the gates close. When the tide recedes, a "head" of water is produced across the barrage and the water is passed over turbines connected to electric generators (Elliot 1996, 231). Small tidal power plants operate around the globe. The largest tidal plant is located in the Rance estuary of Brittany, France. La Rance has a 240 MW capacity, with an average annual output of 480 GWh per year (242).

Energy can also be harnessed from ocean waves as they approach coastal areas. Waves are created indirectly from the solar radiation that drives wind currents. Waves are formed as wind blows across large bodies of water. This energy travels in water, and as it approaches coastal areas, the wavelengths become shorter and the amplitude (or peak height) of the waves is increased (Duckers 1996, 320). Wave energy converters are devices that capture the stored energy in waves and convert it to mechanical energy. They can either be placed perpendicular or parallel to the incident wave front and may be fixed or floating structures. The Aguçadoura wave farm project, the world's first wave power plant, was built off the coast of Portugal in 2006 by Ocean Power Delivery Ltd. (Mellgren 2005).

Wind Energy

Like many of the water energies, wind energy is also formed indirectly from solar energy. Solar radiation causes differential heating and pressure effects to occur in the atmosphere, forming wind currents and weather patterns. The differential heating of landscapes and oceans allows for certain areas in the world to be consistently windy. The kinetic energy of wind can be converted into mechanical power with wind turbines and used to generate electricity. The concept of a wind turbine is the same as that for water or gas turbines, but the design is different in order to exploit the aerodynamic properties of wind. Although there are many different wind turbine designs, two main types are made commercially: horizontal and vertical (whose axis of rotation is vertical).

Significant wind power industries are found in California, Denmark, and the United Kingdom. In California, there are over 15,500 operational wind turbines in the state, with a generating capacity of 16,200 MW. In Denmark, there are over 2,800 operational wind turbines, with a generating capacity of 343 MW. The United Kingdom has been the most recent site for commercial wind energy developments, with over 170 MW of installed wind capacity (Taylor 1996, 304). In addition to large commercial-scale projects, wind power is significantly used in local communities and for small-scale applications.

Biomass Energy

Biomass is the term that is used to describe living matter that is found on the Earth's surface. In terms of energy sources, it refers to the massive amounts of plant (e.g., wood), animal (e.g., dung), and municipal solid waste (MSW) matter that can be used as a fuel to extract useful energy. Like many other renewable sources, biomass energies are indirectly formed from solar energy by photosynthesis, which is the biological process that plants use to convert light energy, carbon, and water into living tissue.

Energy can be extracted from biomass in a variety of ways. Many people rely on direct combustion for the purposes of space heating and cooking. Wood and dung are the most commonly used fuels for these purposes. Thermochemical processing is used to convert biomass to energy. Gasification (when a gaseous fuel is produced from a solid fuel using steam) and pyrolysis (when a material is heated in the absence of air) are two types of thermochemical reactions that are used to produce biofuels with more concentrated energy stores. For example, charcoal is made from wood pyrolysis and coke is made from coal pyrolysis.

Natural processes can also provide fuels for combustion. Anaerobic digestion, which occurs during bacterial decomposition of organic matter in anoxic environments, produces methane. Fermentation is similar to anaerobic digestion, but this process involves organisms that live in oxygenated, or aerobic, environments. Fermentation produces ethanol. These processes can be used to produce fuels from agricultural wastes, municipal solid waste, and even sewage.

Biomass energy represents a significant portion of the renewable energy that is used globally. It is especially important in developing countries where biofuels comprise approximately 35 percent of primary energy sources (Ramage and Scurlock 1996, 139). It is important to highlight the difficulty that exists in measuring biomass consumption. Unlike fossil fuels, which are traded on a global market, many biomass fuels are consumed by people in developing countries who gather their own energy resources. Although it is a significant source for many people around the world, the exact value of its use cannot be quantified (Ramage 1997, 22–23).

Geothermal Energy

Unlike the other forms of renewable energy, geothermal energy is not derived from solar energy. Rather, it arises from heat that exists in the core of the Earth. This heat can be stored in the rocks of the Earth's crust as hot water or in pockets of dry steam. Geothermal energy can either be used to create electricity, or as a direct source of energy for heating. Hydrothermal reservoirs, geopressurized reservoirs, hot dry rock, and magma are the four types of geothermal energy that can be exploited, but the most widely utilized are hydrothermal and hot dry technologies.

Geothermal power plants can utilize both dry steam from hot dry rock reservoirs (vapor that does not contain water) and wet steam from hydrothermal reservoirs, but dry steam is easier to process. A well is drilled into the steam or water reservoir to allow the steam to escape. Once it reaches the surface, dry steam is used to turn turbines for electric generators. If wet steam is extracted, water is separated from the steam at the power plant. This process is called flashing and is employed to protect the turbines from water damage (Brown 1996, 374).

There has been substantial utilization of this resource in the United States, Mexico, and the Philippines. Iceland derives most of its energy from geothermal resources. Globally, over 6 GW of electrical power are produced with geothermal energy, and approximately 4 GW of geothermal power are used annually for domestic heating (Brown 1996, 356).

Despite the variety of renewable energy sources, fossil fuels are consumed far more than any other source. This dependence has many adverse consequences. The next section reveals how fossil fuels came to be the dominant energy resource.

History of Energy Use

The nature and abundance of global energy consumption has drastically changed in the past 150 years. Understanding historical trends and transitions in global energy consumption is important for grasping the complexity of energy use today. This section examines the growth and expansion of usage of the world's energy resources. In particular, it focuses on shifts in primary energy sources, increasing consumption, and the political implications of fossil fuel dependence. First, preindustrial energy consumption and the industrial revolution are discussed, and then, important global events during the twentieth century are examined. Prominent themes in this chapter are the reliance on fossil fuels, the impact of industrialization on energy consumption, and increasing globalization of the energy economy.

Preindustrial Energy Consumption

Throughout most of human history, energy consumption has been relatively low. Human and animal labor provided most of the energy used for agriculture, transportation, and societal growth. Wind, water, and biomass sources were the primary means by which domestic and trade needs were met. This section examines the use of these resources by humans until the 1850s.

Waterwheels were the first devices designed to harness the kinetic energy of flowing water. The first uses of water mills can be traced back to first century BCE, where Romans used them to power grain mills (Smil 1994, 225). Water mills became more common in Europe after 1000 CE. For example, in 1086, it was reported that there were over 5,600 water mills operating in southern and eastern England alone (103). Initially, water mills were used for grain milling, but design innovation and mechanization allowed waterwheels to replace other manual tasks, from papermaking to ore crushing. In the nineteenth century, the waterwheel design was replaced by water turbines, which were more efficient and hence increased power output.

Wind was also an important primary resource. The harnessing of wind energy occurred in the twelfth century in regions of Europe and Asia where water power was not feasible (e.g., in low-lying areas where water heads were nonexistent or in desert areas where water was scarce). The Dutch made vast improvements to windmill design in the 1600s. European use of windmills was by far the greatest in the Netherlands, where in addition to milling grain and pumping water, the Dutch utilized windmills to drain low-lying areas. In the 1800s, the more than 30,000 windmills operating around the North Sea region provided an important source of energy for Europe (Smil 1994, 112).

Biomass energy sources have been extremely valuable to humans throughout history. Wood, dried dung, crop residues, animal oils, and waxes were important for domestic heating, lighting, and food preparation. Additionally, charcoal (the carbon substance produced when wood undergoes pyrolysis) was used for smelting, a process used to purify iron ore (Fe_2O_3). During smelting, high temperatures separate the iron from the oxygen, combining it with carbon to strengthen the alloy. Metallurgy proved to be the most energy-intensive process of the time period. Metal ore needed to be mined, crushed, and then smelted. This final stage required vast amounts of charcoal, and deforestation became a major problem in societies with intense iron trades. By the early 1700s, it is estimated that English iron production required approximately 1,100 square kilometers (approximately 425 square miles) of forest per year to sustain production (Smil 1994, 151). In the 1800s, U.S. iron production required approximately 2,600 square kilometers (approximately 1,004 square miles) of forest (156).

In the seventeenth and eighteenth centuries, deforestation that occurred in England from iron production caused an energy crisis as shortages of fuelwood, lumber, and charcoal increased the prices of these resources. Coal, which was first commercially extracted in Belgium in 1113 and shipped to England as early as 1228, became increasingly used in response to the fuel shortages (Smil 1994, 159). Between 1540 and 1640, most of the coalfields in England were being actively mined. Coke (a carbonized substance produced from the pyrolysis of coal) replaced charcoal as the primary fuel used in metallurgy in the 1700s. The first major energy transition from renewable sources to fossil energies occurred in England during this time.

Industrial Revolution: 1850–1914

The industrial revolution is a broad term used to describe the period in history that marks the rise in manufacturing and industry. During this period, global energy needs dramatically increased and population demographics shifted from rural to urban regions. This section examines energy transitions and energy use during the industrial revolution. Fossil fuels, especially coal, replaced biomass, water, and wood energies as the dominant resource used in society. Technological innovations in engine design and resource extraction allowed industry to become increasingly mechanized and transportation to be revolutionized. Finally, the birth of the oil industry in the 1850s is significant as the beginning of the oil transition.

Coal

In Europe, the transition to coal occurred in the eighteenth century. In the 1700s, European cities were using coal gas as a source for lighting and anthracite for heat. Anthracite was also important in metallurgy as it provided more heat energy than charcoal for the purposes of iron ore smelting. Steam engines were first developed in the late 1600s to increase coal mine production. These engines used either wood or coal combustion to convert the chemical energy of the fuel into mechanical energy. It wasn't until James Watt's innovations in design and efficiency in 1769 that the steam engine became an important part of the industrialized world (Smil 1994, 161). After Watt's patent expired in 1800, a large number of improvements made the steam engine compact, transportable, and efficient. The steam engine powered railways and steamboats, allowing faster transport of goods and people.

In the United States, wood was initially the primary resource that fueled the industrial revolution. The vast amount of forest resources in North America allowed the dependence on biomass energy to continue longer than in Europe. Transportation and domestic heating were the two primary uses of wood in the nineteenth century. In the 1850s, it was estimated that eighteen cords of wood annually were used for home heating (Melosi 1985, 19). (A cord of cut wood is 128 cubic feet and equals a stack that is 4 ft x 4 ft x 8 ft. The energy content of a cord of wood varies from 18,700 MJ/cord for softwood to 30,600 MJ/cord for hardwoods.) Steam engines used approximately 3 million cords of wood per year by the 1830s, and railroads consumed 140 cords per mile per year as late as the 1870s (21). The reliance on wood during the nineteenth century had established a wood-based infrastructure for energy consumption. Industry was designed for charcoal combustion and wood fireplaces dominated space heating applications (23).

Despite the availability of extensive wood resources, coal came to be the dominant fuel that powered the latter half of the industrial revolution in the United States. Anthracite became a vital resource for domestic heating and lighting in urban areas where coal oil and coal gas were cheaper alternatives to wood. Between 1830 and 1850, anthracite coal was used for the smelting of iron. This was especially important during the growth of railways as anthracite allowed ties, rails, and other iron products to be produced more efficiently. Iron smelting became even more efficient with coke made from bituminous coal. Coke replaced anthracite as an industrial fuel in the late 1800s; its use was paralleled by the rise in the steel industry. Innovations in smelting techniques and the availability of coke allowed more efficient removal of impurities and a greater amount of carbon to be forged with the iron alloy.

As an industrial fuel, bituminous coal burned easier and was more compatible with furnace design; however, it did not burn as clean as anthracite, and smoke pollution became a serious problem in urban areas that supported iron and steel industries. While the electric utility industry mainly used anthracite, bituminous resources were utilized during anthracite shortages, causing a brown haze to settle over industrial regions. Smoke pollution in cities, such as Pittsburgh and St. Louis, caused respiratory and public health problems leading to the formation of smoke abatement coalitions, which were important for drawing attention to public health issues associated with energy consumption.

Oil

The Chinese were the first to utilize petroleum products for domestic and commercial purposes. During the Han Dynasty, in 200 BCE, the Chinese used percussion drilling and bamboo pipelines to transport natural gas for the purpose of brine evaporation (Smil 1988, 167). Despite this innovation, the global petroleum industry was not born until the industrial revolution. The distillation of kerosene from petroleum first occurred in 1853 in London by Abraham Gesner. Kerosene was a cheaper illuminant than whale oil, and petroleum became an attractive commodity. In 1859, oil was struck in Titusville, Pennsylvania, by "Colonel" Edwin Drake. The strike caused many oil prospectors to drill in the area. By 1881, Pennsylvania was producing 95 percent of the oil in the United States (Melosi 1985, 39).

In 1863, John D. Rockefeller invested in the new oil industry, purchasing oil refineries in Pennsylvania and Ohio. By 1870, Rockefeller established the Standard Oil Company, which came to be the dominant producer in the oil industry. Standard Oil was effective because it vertically integrated its ventures, operating the drilling, refining, transport, and marketing of oil under the same company. Because of its growth, Standard Oil organized the Standard Oil Trust in 1882. This structure gave the company greater flexibility in managing its business affairs by granting the company's assets to a board of nine individuals (or trustees) who would manage the company's affairs. The arrangement allowed for greater flexibility to control prices in the oil market, granted the company larger tax breaks, and provided a higher return to investors. By 1904, Standard Oil was in control of 90 percent of the kerosene production in the United States; however, the company faced antitrust lawsuits and accusations of operating a monopoly (Melosi 1985, 42). In 1911, a decision by the Supreme Court disbanded the trust into thirty-four separate companies. Three of these companies, Standard Oil of New Jersey (Exxon), Standard Oil of New York (Mobil), and Standard Oil of California (Socal) emerged as dominant players in the global oil industry.

The Pennsylvania boom set off a flurry of oil exploration. Prospects for oil in California began in the 1860s, and other oil fields were discovered in the midwestern and eastern states through the 1890s. The discovery of an extensive oil field in Spindletop, Texas, in 1901 challenged the dominance of the Standard Oil Company. The strike established the oil industry in Texas and gave rise to the Texas Company (Texaco) and Gulf Oil, which became the two other dominant U.S. oil companies in the global market.

In addition to the large U.S. oil companies, two other oil giants emerged during the late 1800s. Royal Dutch Shell and British Petroleum (BP) originated in Western Europe. Shell was founded by Marcus Samuel as a shipping industry that transported coal from Asia in the 1870s. In 1873, Shell gained access to the Russian oil fields in Caucasus. In 1897, the Shell Transport and Trading Company provided intense competition to Standard Oil on the global oil market. At the same time, Royal Dutch, a Netherlands company, began producing oil in the East Indies and was effectively competing against Shell and Standard Oil. In 1906, Shell merged with Royal Dutch.

BP, the other major European oil company, originated in Great Britain. It was founded by William Knox D'Arcy as the Anglo-Persian Company. Having obtained concessions to explore and drill for oil in Persia, it was the first company to exploit oil from the Middle East. Winston Churchill provided a major boost to the Anglo-Persian Company in 1914 when the British government bought half of the company for the purposes of supplying the navy during wartime.

By 1910, these companies, (Standard Oil, Exxon, Mobil, Socal, Texaco, British Petroleum [BP], and Royal Dutch Shell) often called the "Seven Sisters," had established themselves as the dominant participants in the global oil market. During the next fifty years, they expanded their operations throughout the rest of the world. The next section discusses the evolution of the major oil companies and how the world wars of the twentieth century fueled the rise of oil.

Energy, War, and Global Expansion: 1914–1945

In the first half of the twentieth century, the importance of coal in world markets waned and petroleum rose to be the dominant global energy resource. Technology and electricity brought new products to the consumer market, and the mass production of the automobile increased the demand for gasoline. World Wars I and II revealed energy security issues for the United States and Western Europe, and the world emerged from these wars dependent on cheap and abundant supplies of oil. This section addresses important events between 1914 and 1945 that allowed oil to become the dominant resource used in the industrial world and expanded the oil industry into Latin America, Asia, and the Middle East.

World War I began in Europe in 1914. Although the United States did not become an active participant in the conflict until 1917, it aided French and British allies with energy resources and supplies. Coal was still the dominant energy source used in Western Europe and North America, but it was oil that powered motor vehicles, airplanes, and cargo transport during the war. The abundant oil resources from North America allowed the Allies to defeat the Axis powers in 1919 and the United States to emerge from the war as one of the dominant global economic powers.

In contrast to Europe, which sank into a postwar economic depression, the United States experienced a "boom" of consumerism in the 1920s. Cheap gasoline and efficient production lines increased the availability of the automobile. By 1929, 5.6 million cars had been produced in America, roughly one vehicle for every five Americans (Melosi 1985, 108). The availability of cheap electricity also accounted for increasing energy consumption in the United States. Electrical appliances improved the efficiency of domestic tasks, and radios and telephones marked the beginning of the mass communication industry.

After the war, the seven major oil companies expanded their ventures into the Middle East. BP's concessions were already established in Persia (Iran), and Shell had taken part in the establishment of the Turkish Petroleum Company (TPC) in 1912 in Iraq. Shell also had large concessions in Russia. In 1920, Russian oil supplied 15 percent of global consumption, with Shell producing two-thirds of this oil (Sampson 1975, 69). U.S. companies were increasingly left out of Middle Eastern concessions until the 1920s, when the TPC was reorganized to form the Iran Petroleum Company (IPC) and Exxon, Mobil, and Standard Oil gained 23.75 percent of the concessions (Melosi 1985, 107). Two important agreements arose from this alliance. The first was the 1928 Red Line Agreement, which designated an area in the Middle East where IPC companies could not independently seek concessions. This contract discouraged competition from outside companies because much of the area (except for Saudi Arabia) was already controlled by the IPC. It also limited competition between IPC members. The second agreement was the 1928 Achnacarry ("As-Is") Agreement. It created a secret arrangement among the major oil companies to fix oil prices to the Gulf Plus System, which priced oil produced from the participating companies as if it had been produced and shipped from the Gulf of Mexico (Sampson 1975, 73; Melosi 1985, 169). This agreement established the precedent of the petrodollar, where the value of petroleum traded on the world market is measured using the U.S. currency (see chapter 2). The Gulf Plus System protected the market for expensive U.S. oil and allowed Shell and BP to reap large profits from their cheap oil. It was a way for the major oil companies to regulate production among themselves in order to avoid oversupply. It was originally made by Shell, BP, and Exxon, but approved by fifteen other companies including Gulf, Texaco, Socal, and Mobil. The details of this agreement were not fully disclosed until 1952 (Sampson 1975, 73–74).

Although Saudi Arabia contained the largest oil reserves in the world, development of Saudi Arabian oil did not begin until the 1930s. Saudi Arabia was one of the few areas that fell within the Red Line that had not been explored by the major oil companies. In 1933, King Saud, the ruler of Saudi Arabia, granted Socal a concession to explore for oil in the eastern half of the country. Socal and Texaco were not a part of the Red Line Agreement and therefore could seek Middle Eastern concessions independent of the other major oil companies. Socal established the California Arabian Standard Oil Company (CASOC), but was not successful in marketing their resources. In 1936, Texaco joined the concession, and the merger formed the Arabian-American Oil Company (ARAMCO). This marked the first all-U.S. company established in the Middle East.

When World War II started, Socal and Texaco became concerned about the security of their Saudi Arabian concessions. In order to protect the resource, the United States lobbied the British and French governments to release the other major oil companies from the Red Line Agreement. In 1948, BP and Compagnie Francaise de Petroles (CFP, the largest French oil producer) compromised and were given expansions and infrastructure in Iraq in exchange for the dissolution of the Red Line Agreement. With the Red Line erased, Exxon and Mobil joined ARAMCO, with Mobil receiving 10 percent of the concession and the other three companies receiving 30 percent (Sampson 1975, 104).

U.S. oil companies also obtained petroleum concessions in Latin America. World War I increased Mexico's production substantially, but tensions persisting from the Mexican Revolution discouraged investment in Mexico's resources in the 1920s. In

1938, Mexico nationalized its petroleum industry. It created Petromex Petroleos Mexicanos (Pemex) and took possession of the property and oil infrastructure that had been established by seventeen U.S. and European companies. The Mexican expropriation motivated additional tensions between oil companies and the governments of producing countries. Venezuela, which became the leading producer of oil during World War II, demanded greater returns from the oil companies that were operating in their country (namely Shell, Exxon, and Gulf). In order to maintain stability and avoid a trend of nationalism, an agreement was reached in 1948 that provided a fifty-fifty share in all oil profits with the Venezuelan government (Sampson 1975, 109).

World War II demanded high use of energy resources. Oil was important to supply gasoline, jet fuel, and lubricants, while coal was needed for steel manufacturing. The United States maximized production of these resources and provided 80 percent of the oil used by European allies between 1941 and 1945 (Melosi 1985, 181). More efficient transport methods of crude oil were also developed. During the war, the United States constructed two large oil pipelines, the Big Inch, which delivered crude from the Southwestern United States to Pennsylvania, and the Little Big Inch, which stretched from Texas to New Jersey. Although the United States sought to establish a policy of hemispheric solidarity for the purpose of energy security, it recognized the future importance of Middle Eastern oil and made attempts to ensure U.S. concessions in that region. During the next thirty years, the producing countries of the Middle East challenged the power of the Seven Sisters, creating a climate of energy uncertainty.

Middle Eastern Oil: 1945–1970

Oil production and supply became more unstable for the United States and Europe after World War II. Although the United States and Europe remained strong economic powers, they relied increasingly on Middle Eastern oil to supply rising domestic demand. Control of resources by the major oil companies faltered as political tensions halted oil transport and demands for nationalism from producing countries threatened concessions. The development of other energy resources was important during this period, as they provided alternatives to coal and petroleum for electricity generation. This section examines how Middle Eastern tensions and the influence of oil-producing countries shifted the balance of power away from the large multinational companies. Then, it discusses the growth and early development of nuclear power, natural gas, and hydropower resources.

Oil and the Middle East

World War II was devastating to Western Europe. Coal mines that had been under the control of Nazi Germany were damaged and mines in Britain were not able to supply the necessary resources to make up for the shortage. As a result, Europe experienced a large domestic energy crisis. The European Coal Organization (ECO) was formed as the first transnational alliance to respond to an energy crisis. It was responsible for regulating the allocation of coal resources among its ten member states. The member states of the European Coal Organization were Belgium, Denmark, France, Greece, Luxembourg, the Netherlands, Norway, Turkey, the United Kingdom, and the United States. The Eastern European states were left out of the association because of the refusal by the Soviet Union to participate in the ECO (Kapstein 1990, 29). In order to aid its allies, the United States produced the Marshall Plan in 1947, which provided economic and energy aid to Europe by supplying equipment for mine recovery and emphasized a transition to a petroleum-based economy using imports from the Middle East. The Marshall Plan also contained political motivations, as the United States sought to discourage coal imports from Poland, an ally to the Soviet Union. The United States feared that if Poland became economically important to Europe, Communist influence would dominate European politics. In 1947, the ECO was dissolved and the European Coal and Steel Community (ECSC) formed in its place (Hunter and Smith, 2005). The ECSC united European countries and was responsible for responding to energy crises in addition to allocating coal and maintaining markets for coal resources.

The end of World War II was also important for energy dynamics in the United States. In 1947, America became a net importer (rather than a net exporter) of oil. This shift was due to increasing consumption of oil, insufficient means for controlling waste during the production of petroleum, and the large use of oil during the war. It is a significant point in energy history because it marks the beginning of the United States' dependence on foreign oil. Although the United States still supplied a significant amount of oil to the energy market, its influence in the energy economy increasingly relied on the ability of multinational companies to secure international oil resources. For the next thirty years, this reliance proved to be difficult as political tensions hindered the flow of oil from the Middle East and producing countries became more powerful.

Following the nationalization and profit-sharing success that occurred in Latin America, Middle Eastern producing nations became interested in examining their own relationships with oil companies. From 1951 to 1954, Iran was the first country to attempt nationalization of its oil reserves. In 1951, after profit-sharing negotiations failed with the Anglo-Iranian Oil Company (AIOC, previously known as the Iran Petroleum Company), the AIOC refinery complex was shut down. The Iranian nationalization was not completely successful. Iran was not able to secure contracts with other oil companies, and overproduction in world oil allowed supplies from other producing countries to supplement the market. In 1954, an agreement was reached between Iran, the United States, and Great Britain that provided for national ownership of all AIOC properties by the National Iranian Oil Company (NIOC), but relied on a consortium of foreign oil companies to produce the oil.

Although the energy market emerged from the Iranian crisis relatively unscathed, the Suez Crisis in 1955 proved to be more devastating to European energy security. By 1956, oil accounted for approximately 22 percent of total European energy consumption with 90 percent of this oil being supplied by the Middle East and 70 percent shipped through the Suez Canal (a man-made waterway that was operated by the British-owned Suez Canal Company) (Kapstein 1990, 103–105). In July 1956, the Egyptian president, Gamal Abdel Nassar, nationalized the Suez Canal. Great Britain, France, and Israel, fearful that the oil supply would be disrupted, responded by organizing a coordinated attack against Egypt (without the support of the United States) for the purpose of taking control of the canal. Although the fighting ceased by December 1956, the canal remained closed to oil shipments until May 1957 and an extensive energy crisis plagued Europe. The closing of the canal shut off two-thirds of the oil shipped to Europe. Although emergency supplies flowed from the United States, this nation was reluctant to provide support and Europe realized the consequences of its reliance on Middle Eastern oil. In response to the crisis, the Organization for European Economic Cooperation (OEEC) developed energy strategies for future shortages that created emergency petroleum stockpiles and diversified member countries' oil resources by securing reserves in North Africa and Russia.

The Suez Crisis also impacted energy policy in the United States and raised concern over increasing reliance on foreign sources. In 1959, President Dwight Eisenhower established the Mandatory Oil Import Program (MOIP), which limited petroleum imports at a set amount and controlled them with the issuance of "quota tickets" to individual companies. Although it was a national policy constructed to address energy security issues in the United States, the MOIP affected the global energy market by lowering the price of international oil. Because of the quotas, demand for foreign oil was lowered in the United States, but world supply of oil remained the same. This imbalance created a trend of decreasing oil prices. Additionally, the opening of new oil markets in northern Africa (Algeria) increased the supply of world oil and caused oil companies to implement two successive price reductions in 1959 and 1960. In light of the declining world oil prices, major producing countries faced declining revenue and in response formed the Organization of Petroleum Exporting Countries (OPEC) in 1960. Members of OPEC (Saudi Arabia, Iran, Iraq, Kuwait, and Venezuela) required that original prices be restored and demanded consultation prior to price reductions. OPEC's influence was minor throughout the 1960s, but the cartel contributed to the increasing unease that was felt by consuming nations. In response to OPEC, major industrialized nations established the Organization for Economic Cooperation and Development (OECD) in 1961. This organization replaced the OEEC and extended membership to the United States, Canada, Japan, New Zealand, and Australia. Although not established to deal with energy matters exclusively, the OECD in part developed strategies to deal with energy shortages among industrialized nations.

The energy emergency strategies that were developed after the Suez Crisis helped alleviate the impacts of another oil crisis that emerged in 1967 when tensions escalated between Israel and Arab nations. In June 1967, Israel preemptively attacked Egypt, marking the beginning of the Six-Day War. Because of their support for Israel, Arab states implemented an oil embargo against the United States and Europe. The oil shortage that followed was alleviated by an increase in exports from Venezuela and Iran (which did not participate in the embargo) and from an increase in U.S. production. Although the embargo was lifted by the end of July, unrest intensified among Arab producing nations over Western support for Israel. In 1968, Arab states founded the Organization for Arab Petroleum Exporting Countries (OAPEC) for the purpose of uniting political interests in the Arab nations. The conflict between Arab and Israeli nations was heightened during the seventies, leading to energy crises that altered the global energy economy.

Natural Gas, Hydropower, and Nuclear Energy

Natural gas, hydropower, and nuclear energy are three primary sources that became important during the twentieth century. Natural gas resources are generally found associated with oil reserves, but it wasn't until the 1920s, when advances in pipeline design allowed for easier transport, that gas became a commodity. Natural gas burned cleaner than oil or coal and hence was attractive as a domestic fuel. In the United States, an extensive gas pipeline network was constructed following the passage in 1938 of the Natural Gas Act, which regulated the price of natural gas between producing states and consuming states. In Russia, discovery and expansion of natural gas fields in Siberia, the Ukraine, and North Caucasus in the 1950s and 1960s led to the development of an extensive natural gas pipeline through central Russia (Dienes and Shabad 1979, 75). The development of natural gas pipelines in Russia also allowed Europe to import natural gas. European gas consumption increased with the discovery of gas and oil reserves in the North Sea in the 1960s.

Hydropower also came to be an important energy resource for the generation of electricity. Governments in industrialized countries sponsored large water projects. The United States constructed large dams along many rivers and developed river basin co-operations to coordinate the distribution of electricity. One such project, the Tennessee Valley Authority, was created in the 1930s. In the Soviet Union, development of large-scale hydropower projects also began in the 1930s along the Dnieper River, and hydropower developments expanded drastically in the 1950s into Siberia and Central Russia (Dienes and Shabad 1979, 136–137). Hydropower provided a cheap, efficient resource for electricity and irrigation. Large water projects were extended into many developing countries, such as Brazil and India, throughout the 1960s and 1970s. Worldwide, construction of large dams peaked in the mid-1960s, with approximately 1,000 large dams being constructed per year (Khagram 2004, 8).

Nuclear power emerged in the United States with the development and use of the atomic bomb during World War II. After the war, nuclear energy's promise as an alternative to polluting coal made it an attractive energy source to the power industry. In 1953, Eisenhower's famous "Atoms for Peace" speech at the United Nations pledged support for the peaceful development of nuclear technology. This speech led to the establishment of the International Atomic Energy Agency (IAEA) in 1956 with eighty-one member countries. Its goal was to act as a watchdog agency, providing verification of the safety and security of nuclear development around the world (Fischer 1997, 1).

In the United States, the federal government passed the Atomic Energy Act in 1946, which established the Atomic Energy Commission (AEC) and the Joint Committee on Atomic Energy (JCAE) in Congress. These two governing bodies regulated the nuclear power industry and established federal ownership of nuclear fuels. The 1954 Atomic Energy Act and the 1957 Price-Anderson Act provided incentives for private nuclear power development by allowing private firms to own nuclear reactors and by limiting the liability of potential nuclear accidents with subsidies that would cover the damages. Although these measures increased the development of nuclear resources, public health and safety concerns hindered the expansion of nuclear power.

The nuclear power industry was also adopted in other parts of the world. In Russia, the first nuclear reactor became operational in 1954. By 1975, more than 6,200 MW of generating capacity was installed (Pryde 1979, 151). In Europe, nuclear energy offered an alternative to coal and a buffer against the volatility of the oil market in the 1960s and 1970s. Coordination of nuclear technology and resources was examined by Euratom, an organization established by the ECSC in 1955 to research nuclear technology development in European nations. Euratom's influence grew after the Suez Crisis, but multinational coordination was ultimately undermined by initiatives in member countries to develop nuclear resources independently. This was the case in the 1970s when the security of energy resources was increasingly threatened.

Energy Crisis: 1970–1980

The energy crisis that emerged during the 1970s resulted from increasing tensions in the Middle East and a growing concern for environmental degradation from energy use. Conflict between Israel and the Arab world intensified, creating drastic consequences for the energy market. Additionally, increasing apprehension over the environmental impacts of energy production and consumption altered the regulatory climate in which the energy industry operated. This section describes the energy crises that emerged from the 1973 oil embargo and how these events changed the energy industry. The environmental effects of energy consumption are examined further in chapter 2.

Oil Embargo

Until 1970, impacts to the energy market as a result of Middle East tensions were buffered by the capacity of the major oil companies to control production and prices in other parts of the world. Major consuming nations also cushioned the impact to world markets through cooperation and support. This alliance changed in the 1970s as allied nations sought to protect their own energy interests over global energy stability (Kapstein 1990, 152).

Throughout the 1960s, the Middle East became the center of world oil, accounting for 38 percent of world production and 90 percent of international trade (Melosi 1985, 249). The Yom Kippur War of 1973 provided the catalyst for OPEC's emergence as the leader in oil production and pricing. When Egypt and Syria launched a surprise attack on Israel in October 1973, the United States airlifted weapons to aid the Israeli defense. In response, members of OAPEC initiated an oil embargo against the United States and other allies of Israel. Saudi Arabia participated in the embargo, and ARAMCO was ordered to cut production by 25 percent and cease shipments to the United States. Additionally, OPEC increased world prices of oil from $3.00 per barrel to $11.65 per barrel (Hatch 1986, 32). Although the major oil-producing companies responded by increasing production in other countries, shipments of crude to the United States dropped from 6 million to 5 million barrels per day, and the country experienced the largest energy crisis in its history (Melosi 1985, 238).

The embargo ended in March 1974, six months after it had been initiated. It was largely aimed at the United States and allies of Israel, but the action by OPEC had produced drastic price increases in crude oil. Although a global recession resulted in a decrease in demand for OPEC oil, the cartel maintained its pricing power and the oil market remained relatively stable between 1974 and 1978. In 1978, tensions in the Middle East again demonstrated the volatility of the oil market when the Iranian Revolution raised the price of oil on the spot market to as high as $45 per barrel (Melosi 1985, 282).

Global Response to the Embargo

The oil embargo of 1973–1974 not only demonstrated the consequences of dependence on foreign oil, it also showed the developed world that it could not always rely on cheap, abundant oil. In 1974, the International Energy Agency (IEA) was established within the framework of the OECD as a direct response to the embargo. Its mission was to develop strategies for energy security during emergencies and to reduce member countries' dependence on oil (IEA 2005). Non-OECD countries also developed strategies for oil emergencies. The Association for Southeast Asian Nations (ASEAN) developed a Council on Petroleum (AS-COPE) to coordinate the management of energy resources in member countries in the event of shortages or oversupply of petroleum (Karki, Mann, and Salehfar 2005, 499). (The ASEAN was established in 1967. It is made up of ten countries: Indonesia, Malaysia, Philippines, Singapore, Thailand, Brunei Darussalam, Vietnam, Laos People's Republic, Myanmar [formerly Burma], and Cambodia.)

Many countries implemented national energy policies to address concern over energy security. In Europe, France, Germany, and the Netherlands began to expand their nuclear programs. Great Britain continued to diversify its oil imports, exploiting resources in the North Sea and Algeria. Europe also began to increase oil imports from the Soviet Union. In the United States, the embargo spurred development of domestic supplies as well as the formation of a unified energy policy and conservation measures. U.S. energy policy is discussed further in chapter 3.

Despite the many actions taken to alleviate concerns over oil security, no significant measures were taken to move energy dependence toward renewable sources. Consequently, global energy use has become more reliant upon fossil fuels in the thirty years since the oil embargo. In addition, the use of these sources has rapidly accelerated with the industrialization of many developing countries. In the next chapter, events associated with global energy use are described in the context of the social and environmental problems they create.

Conclusion

This chapter has provided an overview of energy fundamentals and historical energy trends and transitions. Major themes of the chapter focus on the nature of energy, the sources of energy, and the importance of energy to society. The historical account of energy transitions highlights the dominance of fossil fuels throughout the past 150 years and the importance of energy stability for national security. As the world becomes more industrialized, the demand for energy is going to dramatically increase. The reliance on fossil energies is problematic for nations as they seek to secure adequate supplies of energy resources and for the world as it battles environmental and social problems associated with the trends in energy use. The next chapter describes these problems.

References

Alexander, G. 1996. "An Overview: The Context of Renewable Energy Technologies." In *Renewable Energy: Power for a Sustainable Future,* edited by G. Boyle, 1–39. Oxford, England: Oxford University Press.

Boyle, G. 1996. "Solar Photovoltaics." In *Renewable Energy: Power for a Sustainable Future,* edited by G. Boyle, 89–136. Oxford, England: Oxford University Press.

Brown, G. 1996. "Geothermal Energy." In *Renewable Energy: Power for a Sustainable Future,* edited by G. Boyle, 353–392. Oxford, England: Oxford University Press.

Dienes, L., and T. Shabad. 1979. *The Soviet Energy System: Resource Use and Policies.* New York: John Wiley & Sons.

Duckers, L. 1996. "Wave Energy." In *Renewable Energy: Power for a Sustainable Future,* edited by G. Boyle, 321–352. Oxford, England: Oxford University Press.

Elliot, D. 1996. "Tidal Power." In *Renewable Energy: Power for a Sustainable Future,* edited by G. Boyle, 332–370. Oxford, England: Oxford University Press.

Energy Information Administration (EIA). 2004a. Energy Kids. http://www.eia.doe.gov/kids/energyfacts/index.html (accessed July 7, 2005).

Energy Information Administration (EIA). 2004b. "International Energy." In *Annual Energy Review.* http://www.eia.doe.gov/emeu/aer/contents.html (accessed June 30, 2005).

Energy Information Administration (EIA). 2004c. "World Consumption of Primary Energy by Energy Type and Selected Country Groups (Standard Units), 1980–2002." *International Energy Annual 2002.* http://www.eia.doe.gov/iea/ (accessed June 29, 2005).

Energy Information Administration (EIA). 2005a. *International Energy Annual 2003 Glossary.* May 25, 2005. http://www.eia.doe.gov/emeu/iea/glossary.html (accessed July 7, 2005).

Energy Information Administration (EIA). 2005b. "World Proved Reserves of Oil and Natural Gas, Most Recent Estimates." http://www.eia.doe.gov/iea/ (accessed June 29, 2005).

Everett, B. 1996. "Solar Thermal Energy." In *Renewable Energy: Power for a Sustainable Future,* edited by G. Boyle, 41–88. Oxford, England: Oxford University Press.

Fischer, D. 1997. *History of the IAEA: The First Forty Years.* Vienna: International Atomic Energy Agency.

Hatch, M. T. 1986. *Politics and Nuclear Power: Energy Policy in Western Europe.* Lexington: University Press of Kentucky.

Hunter, J., and Z. A. Smith. 2005. *Protecting Our Environment: Lessons from the European Union.* Albany: State University of New York Press.

Ingersoll, J. G. 1990. "Solar Thermal Energy." In *The Energy Sourcebook: A Guide to Resources, Technology and Policy,* edited by R. Howes and A. Fainberg. New York: American Institute of Physics.

International Energy Agency (IEA). 2005. "An Overview." http://www.iea.org/ (accessed July 21, 2005).

Kapstein, E. B. 1990. *The Insecure Alliance: Energy Crises and Western Politics since 1944.* New York: Oxford University Press.

Karki, S. K., M. D. Mann, and H. Salehfar. 2005. "Energy and Environment in the ASEAN: Challenges and Opportunities." *Energy Policy* 33, 4 (March 2005): 499–509.

Khagram, Sanjeev. 2004. *Dams and Development: Transnational Struggles for Water and Power.* Ithaca, NY: Cornell University Press.

Mellgren, D. 2005. "Groundbreaking Wave Power Electricity Project to Be Built off Portugal." *Associated Press Environmental News Network.* May 23, 2005. http://www.enn.com/today.html?id=7791 (accessed July 7, 2005).

Melosi, M. V. 1985. *Coping with Abundance: Energy and Environment in Industrial America.* Philadelphia: Temple University Press.

Miller, E. W., and R. M. Miller. 1993. *Energy and American Society: A Reference Handbook*. Santa Barbara, CA: ABC-CLIO.

Nuclear Energy Agency (NEA). 2005. "Facts and Figures." *Nuclear Energy Data 2005*. http://www.nea.fr/html/general/facts.html (accessed August 3, 2005).

Pyrde, P. R. 1979. "Nuclear Power." In *The Soviet Energy System: Resource Use and Policies*, edited by L. Dienes and T. Shabad, 151–182. New York: John Wiley & Sons.

Ramage, J. 1996. "Hydroelectricity." In *Renewable Energy: Power for a Sustainable Future*, edited by G. Boyle, 181–226. Oxford, England: Oxford University Press.

Ramage, J. 1997. *Energy: A Guidebook, New Edition*. Oxford, England: Oxford University Press.

Ramage, J., and J. Scurlock. 1996. "Biomass." In *Renewable Energy: Power for a Sustainable Future*, edited by G. Boyle, 137–182. Oxford, England: Oxford University Press.

Rose, D. J. 1986. *Learning about Energy*. New York: Plenum Press.

Sampson, A. 1975. *The Seven Sisters: The Great Oil Companies and the World They Shaped*. New York: Viking Press.

Smil, V. 1988. *Energy in China's Modernization*. Armonk, NY: M. E. Sharpe.

Smil, V. 1994. *Energy in World History*. Boulder, CO: Westview Press.

Smil, V. 1999. *Energies: An Illustrated Guide to the Biosphere and Civilization*. Cambridge, MA: MIT Press.

Smil, V. 2003. *Energy at the Crossroads: Global Perspectives and Uncertainties*. Cambridge, MA: MIT Press.

Stoker, H. S., S. L. Seager, and R. L. Capener. 1975. *Energy: From Source to Use*. Glenview, IL: Scott, Foresman and Company.

Taylor, D. 1996. "Wind Energy." In *Renewable Energy: Power for a Sustainable Future*, edited by G. Boyle, 270-320. Oxford, England: Oxford University Press.

Wolfson, R., and J. M. Pasachoff. 1995. *Physics: With Modern Physics for Scientists and Engineers*. 2nd ed. New York: HarperCollins College Publishers.

2

Problems, Controversies, and Solutions

Introduction

Chapter 1 described how energy is a vital, fundamental component of society. Society depends on energy availability and distribution to function. It showed how humans have harnessed and used energy from different sources and how the primary sources of energy have changed over time.

This chapter discusses trends that impact worldwide energy use, the global environmental and social problems associated with energy consumption, and potential solutions to problems. The first section describes how energy systems and economics are intertwined and develops the context necessary for understanding global energy issues. The second section examines environmental and social problems of energy use. Finally, potential solutions to energy problems are discussed.

Energy and Economics

Energy is fundamental to economic development. Without energy resources, businesses would not be able to light their stores, people and goods would not be able to reach markets on the other side of the world, homes and schools would be more difficult to heat, and manufacturing sectors would not be able to produce clothes, radios, or any products used on a daily basis. In short, the world economy depends on energy, and the largest economies of the world (i.e., those of industrialized countries) rely on cheap and abundant supplies of energy. This reliance has large social and environmental costs. In order to understand those costs, one must first understand the link between energy and the economy.

Energy use in society is often correlated with the gross domestic product (GDP) of national economies. GDP is simply a measure of the goods and services produced annually in a particular country. For example, according to the World Bank, the GDP of the United States was $11.667 trillion in 2004 (World Bank 2005), making it the largest in the world. GDP is often used in discussions of energy use because there is a strong linear correlation between the amount of energy consumed in a society per capita (per person, per year) and a nation's GDP (Smil 2003, 68). That is, in general if a country's GDP is higher relative to other countries', more energy is consumed per capita in that country (figure 2.1). This link is often described using a measure called energy intensity (EI), which is the ratio of energy use in common units of energy (e.g., gigajoules [GJ]) to GDP (e.g., measured in dollars).

EI = energy consumed (GJ) / GDP per capita (U.S. dollars)

So if a country's EI value is relatively low, then it is not thought to consume a lot of energy and it might have a lower GDP. Conversely, a high EI would indicate that both energy consumption and GDP are high.

It is important to note that the relationship between energy consumption and GDP is not always as clear as the simple linear model described above. Because of this, EI by itself cannot be used to make sweeping claims about energy consumption. A country may have a relatively high GDP and a low EI if its industries are more efficient at producing and consuming energy resources. Recall that energy efficiency was a term introduced in the previous chapter as being the ratio of useful energy output to total energy input. At a societal level, the idea of energy efficiency can be thought of as how much energy services are gained (i.e., lighting, transportation, etc.) from the same energy inputs (i.e., coal, oil, etc.) (Smil 2003, 318). Efficiency in energy use depends upon what energy resources are used and the technology used in energy conversions. For example, hydropower plants are 90 percent efficient in converting stored potential energy into electricity,

FIGURE 2.1
A General Correlation between GDP and Energy Consumption

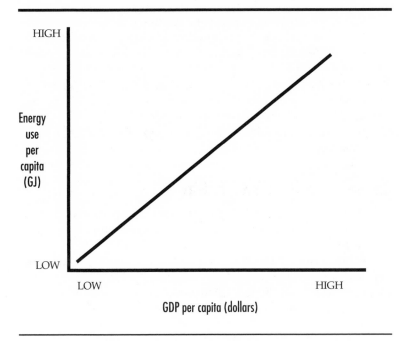

whereas coal-fired power plants are between 30 and 40 percent efficient depending on what combustion technology is used (Manahan 2000). Because energy efficiencies are different for all countries, it is important to understand that GDP may not always be the best indicator of energy use in a society and vice versa. While energy intensities are a common descriptive measure used to understand energy dynamics, they should be considered within the context of what a country's primary energy resources are and the efficiency of its infrastructure. This is also true for linking energy use to quality of life measures. Analysts often assume that high energy use in society correlates with a higher quality of life. While this may be the general case, energy dynamics need to be considered in context.

Despite these difficulties, broad observations regarding EI trends over time and economic development have been useful for predicting future energy trends. In general, energy intensities increase as a country begins to industrialize (Smil 2003, 68). That is, as a country starts to produce more in manufacturing (e.g., textiles) and industrial (e.g., steel) sectors, it uses more energy. GDP increases because more goods and services are being produced, but because the energy infrastructure is not developed compared to its industrialized counterparts, energy is being used less efficiently. After a relatively short peak, a country's EI drops as its energy systems become more efficient (figure 2.2). This particular correlation is important in examining global energy trends as well as other significant aspects of global energy use and economics.

Energy Markets and Pricing

The global market for energy is the means by which energy resources are traded among countries and within entities that supply, produce, and distribute energy. Most nations do not have domestic natural resources available to support their energy needs, so they buy energy from other countries. These transactions are most often carried out by large energy companies. Transactions are complex and constantly changing to reflect variable energy prices and investor attitudes. Energy giants, such as the oil companies of BP, Shell, and Exxon-Mobil, dominate the global energy markets. Oil is not the only thing sold by energy companies, which make an incredible amount of money selling energy resources and services. It is important to consider the power wielded by energy giants when thinking about the problems associated with global energy dynamics. This section describes why energy pricing is important and how governments work to stabilize prices.

Energy prices are important because cheap energy allows for economic growth. Nations strive for continuous economic growth because it increases the quality of life for citizens and provides greater trade opportunities. High energy prices and market volatility make it difficult for some countries to obtain the necessary resources. Hence, it is in a nation's best interest to keep energy prices stable.

The most common way that prices are manipulated by governments is through energy subsidies. These are payments or rewards granted by governments to energy companies for the purpose of minimizing the cost of energy produced for public con-

FIGURE 2.2
General Trends in Energy Intensity during Industrial Development

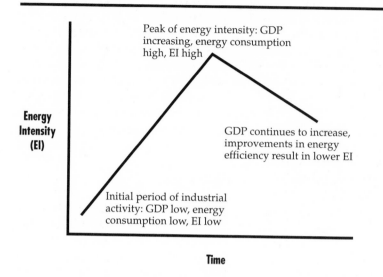

Peak of energy intensity: GDP increasing, energy consumption high, EI high

Energy Intensity (EI)

GDP continues to increase, improvements in energy efficiency result in lower EI

Initial period of industrial activity: GDP low, energy consumption low, EI low

Time

sumption. They are important components of national energy policies because they are thought to promote economic growth. Subsidies can be in the form of direct payments, tax exemptions, or funding for research and development that may provide potential energy sources in the future (EIA 1999). Since energy prices, particularly for crude oil, are determined from fluctuations in demand and supply curves, subsidies affect prices because they encourage a supply increase in a particular energy resource on the market. By subsidizing certain resources, a government can influence which energy sources are utilized in society.

The main problem that arises with subsidies is that it makes it more difficult to determine real energy prices based on costs of production. Because of subsidies, people do not pay the real price that it costs to produce the resource. Furthermore, the largest subsidies go to energy sources that are damaging to the environment. As a result, pricing schemes are distorted because they do not reflect the environmental and health costs associated with energy production. Another aspect of energy subsidies is that they encourage wasteful energy consumption. When energy is cheap, people aren't concerned about conserving resources. Moreover, energy companies often promote excessive consumption because they receive larger subsidies if they can supply more energy. Proposed solutions to these issues are discussed a little later.

An interesting point to note regarding energy markets is that the most traded energy commodity in the world today is crude oil. About 65 percent of the oil that is produced globally is exported to more than 130 countries (Smil 2003, 45). Approximately 85 million barrels of oil per day were traded on the market in December 2005 at an average price of $59.45 per barrel (IEA 2006). As of this publication, the only accepted currency for oil trade is the U.S. dollar. This means that countries must have U.S. currency in their treasuries or banks to purchase crude oil; they cannot use their own national currencies. (The "petrodollar" was established with the "As-Is" Agreement [see chapter 1].) The use of petrodollars is particularly advantageous to the United States, as it gives U.S. companies dominance in the oil markets. However, it is inevitable that this standard will be challenged as markets become increasingly globalized.

Globalization

More and more, energy markets are becoming increasingly globalized as companies are becoming more integrated within different regions and across national boundaries. As information processing and transportation rapidly expand across the globe, access to energy resources becomes more available for developing countries. Additionally, developing countries that have abundant natural resources (i.e., the raw ingredients for energy conversions) have increasing opportunities to access energy markets and profit from the sale of oil or natural gas.

Market liberalization and privatization are important characteristics of globalization. Globalization has increased the power of transnational corporations (TNCs). TNCs are companies that own industries, operations, and distribution centers outside of their parent country's borders. As TNCs become more powerful, national companies (those that produce and distribute goods within the borders of their parent countries) are pressured to sell their assets to TNCs. This process is market liberalization. Market privatization occurs when energy companies owned by national governments (i.e., state owned) sell their resources to private companies and no longer control resource extraction. On one hand, privatization gives developing countries a means to enter the global market. On the other hand, it causes many developing countries to lose access to their own natural resources. Critics contend that these countries are being exploited for their oil and gas reserves.

All in all, there are positive and negative aspects with respect to globalization. Globalization has the potential to raise the standard of living in many impoverished countries. However, in the absence of global environmental norms and economic accountability, this trend can be damaging to both the environments and the economies of developing countries. Furthermore, high foreign investment in fossil energies can create a reliance on profits that are characteristically subject to unstable price shifts. If a large percentage of a country's revenue is based on an unstable oil market, its economy can experience boom and bust cycles, where cheap prices encourage investment, but rapid price shifts can lead to devastating recessions. These uncertain economic patterns can increase a nation's debt and exacerbate poverty (Clapp and Dauvergne 2005, 172).

Often developing countries do not have the resources to adopt and enforce environmental standards. If transnational energy companies invest in these countries, they may not develop the resources according to the environmental standards of their parent country. Without regulation, oil and natural gas development can significantly degrade the environment, reduce the amount of arable land, and pollute the water and air. Such was the case in Nigeria, Africa's largest oil producer, where the oil giant Chevron developed an incredible oil infrastructure and pipeline system throughout the Niger Delta. Because Chevron did not need to adhere to environmental laws, the safety of the Nigerian people was at risk. For example, frequent gas flares (a practice used to cheaply remove unwanted gas from the oil) endangered people who lived close to the pipelines. Often the flares caused explosions, like one in 2000 that killed more than 700 people (Goodman and Goodman 2004, 71). Flares also released large amounts of pollution, endangering human health and deteriorating ecosystems. Despite these risks endured by millions of Nigerians, many still live in desperate poverty, unable to afford the basic energy resources that are exported from their own country.

The case of Nigeria provides an example of how globalization can be important in understanding energy use, but what about some of the other economic concepts? Why are energy intensities relevant? What is the importance of a global energy market? And why must governments carefully plan how they subsidize energy? The next section demonstrates how these concepts are important to trends in energy use.

Energy Trends

The linkage of energy to the global economy can only be put into perspective with an analysis of events and trends highlighting how these components of society are intertwined. The following discussion seeks to deepen understanding of fundamental energy and economy concepts using real examples and demonstrates how energy planning has political and economic implications in a country's development. First is an examination of how highly subsidized and state-centralized energy systems of the former Soviet Union resulted in incredible energy inefficiencies, causing former Soviet republics to be largely uncompetitive in a capitalist, global market. Next is a discussion of the rapid rate of industrialization occurring in the developing world and how this trend is important to consider in the context of depleting fossil fuel reserves.

Centralized Energy Systems

The collapse of the Soviet Union in 1989 is an important example of how economies are linked to energy systems. The Communist regime developed many energy-intensive industries with little regard to environmental effects. The Soviet republics pursued many years of rapid economic growth without developing means to increase energy efficiency. Energy resources were state-owned and heavily subsidized by the government to promote economic growth. This promotion led to excessive consumption without regard to conservation. Because measures were never taken to improve energy efficiencies, energy intensities in the region remained very high. In other words, despite producing only a fraction of the GDP that was produced by the European Union (EU), energy consumption in the republics was many times greater.

Even though it has been over twenty-five years since the collapse of the Soviet Union, the energy systems in former Soviet bloc countries remain inefficient. In addition, ecological damage that ensued as a result of grossly inefficient infrastructure poses a problem for many of the countries. During the Communist era, electricity generation relied mainly on coal-fired power plants. These facilities often did not employ the efficient designs developed since the 1970s and pollution control measures were largely absent. As a result, some countries have had to contend with polluted waters and lower crop yields.

This example is important because it demonstrates how poorly planned energy systems can cause long-term problems. In order for former Soviet countries to become competitive in a global, capitalist market, it is necessary for them to improve energy efficiencies. The transition is difficult because it requires large amounts of capital funding, which is lacking in countries whose economies have been devastated from the fall of their centralized economic and political systems. Hence, their assimilation into capitalist economies has been difficult.

Industrialization of Developing Countries

World energy demand continues to rise as more developing countries become industrialized. Increasing along with demand is the scarcity of fossil fuel resources and the consequences of fossil fuel reliance. Asia is the area of the world that is expected to experience the largest increase in energy demand over the next twenty years. This region, with over 50 percent of the global population, is expected to surpass the United States and Europe as the world's largest consumer of primary energy sources by 2010 (Manning 2000, 59). In a globalized economy, energy trends in Asia are important for future global energy security. Economic growth in Asia has rapidly increased over the last forty years, rising from 4 percent of world GDP in 1960 to 25 percent in 1995 (62). Much of the resulting increase in energy demand has occurred in China.

China began its industrialization in the 1940s and 1950s during a time when it supplied not only its own initial growth but provided coal resources to Stalinist Russia (Smil 1998, 85). China, endowed with abundant coal reserves, relies on coal to supply approximately 68 percent of its energy needs, and petroleum resources account for 25 percent of energy consumption (Manning 2000, 93). Chinese demand for oil has grown, and the country is expected to become the world's largest importer of oil by 2025 (105). This trend creates substantial impacts for the oil industry as Chinese petroleum companies increase their influence in the Middle East. In 1997, the Chinese National Petroleum Company

(CNPC) outbid other major oil companies (including Exxon-Mobil) to acquire oil fields in Kazakhstan and develop petroleum resources in Iran, Iraq, Venezuela, and Sudan (Manning 2000, 86). In 2005, the Chinese oil company Cnooc sought to purchase the U.S. oil giant Unocal. Cnooc's $18.5 billion offer (the highest bid) for the company resulted in heightened tensions between the United States and China (Barboza and Sorkin 2005). Many U.S. businesses and lawmakers protested the potential acquisition, and in the face of mounting political opposition, Cnooc withdrew its offer. Although Cnooc was not successful in acquiring Unocal, the attempt demonstrates the growing Chinese presence in the global market and potential threat to U.S. energy security because of increased competition for Middle East oil resources. Chinese energy companies are becoming increasingly globalized as they seek ways to fuel their own economic growth via acquisitions of foreign resources and companies. As China's power grows, it could threaten the dominance of U.S. companies in the oil market, resulting in strained political relations between the two countries. Since China is the primary trading partner of the United States, heightened tensions between the two countries could be detrimental to other business relations. (Conflicts associated with energy are discussed in greater detail later on.)

Increased energy consumption in China reflects a larger trend of energy demand emerging from Asia. Energy consumption in India, South Korea, and Indonesia has drastically increased, and Japan continues to import the majority of its energy resources. Although it is inevitable that these and other nations become industrialized, it is important to remember that the energy resources used to support this industrialization are nonrenewable fuels (coal, petroleum, natural gas, and nuclear). In China, 97.5 percent of energy is derived from these resources (Manning 2000, 93). In India, they account for 92 percent (125). (These numbers do not include estimates of biomass fuels. Biomass resources are especially important in rural areas of China and India where they are used for domestic heating and food preparation purposes.)

These trends are particularly relevant in light of the earlier discussion regarding economic development and energy intensities. While the EI in China has been decreasing steadily since 1980 (Geller 2003, 96), EIs in many other developing countries are rapidly increasing as they use more energy in a less efficient way than the developed countries of Western Europe and North America. At some point, the EIs in these countries are expected to peak and then decrease. But until that happens, energy consumption is greater than what it would be if more efficient systems were being utilized. Inefficient consumption has both environmental and social implications. The next section discusses why growth in the energy sector is especially relevant in the context of the problems that arise from dependence on fossil energies.

Environmental and Social Problems

Why do trends in global energy use matter? Economic growth is good for developing countries; it creates jobs, increases trade, and helps pull people out of poverty, right? Overall, it's good when countries experience economic growth, so why are people concerned with the increasing demands for energy? These questions are answered in the next couple of sections.

There are many social and environmental problems related to the world's reliance on commercial fossil energies. The processes of extracting, refining, distributing, and using energy create many problems, including air and water pollution and land degradation. These impacts may have serious consequences for human health and well-being. In addition to environmental effects, current global energy dynamics result in an extreme imbalance of energy distribution. While industrialized countries enjoy the luxuries of motorized vehicles, TVs, and kitchen appliances, many developing countries cannot even supply the basic energy services needed for cooking, heating, lighting, and sanitation. As a result, the quality of life for people in those parts of the world remains very low. Projections of global energy use cause concern that these problems may become exacerbated. This section first examines the environmental problems that result from energy use and then discusses the social problems associated with the energy use trends.

Environmental Problems

The consumption of energy resources inevitably compromises environmental quality. There are no energy sources that are completely environmentally friendly, but some are more damaging than others. Air pollution, water pollution, and land degradation are a result of the extraction, transportation, processing, and combustion of fossil fuels. Additionally, combusted fossil energies release greenhouse gases that impact climate change. Nuclear energy, while it does not produce air pollution or greenhouse gases, presents its own unique public health and environmental issues. Even renewable energy resources have problems associated with their use. This section first examines general problems caused by fossil fuels and their consequences to human and ecosystem health. Next, it discusses particularly harmful effects characteristic to the nuclear fuel cycle. A fuel cycle is a term that refers to the processes of extraction, transportation, processing, and consumption for each fuel type. Finally, it briefly discusses environmental impacts that occur from the use of renewable energies.

Production and Transportation (Fossil Energies)

Mining for coal and drilling for oil and natural gas have large environmental costs associated with them. Coal mining, which is a very dangerous occupation, is devastating to the land. Coal can be mined in two ways: deep shaft mining and surface mining. Deep shaft mining is used in areas where coal seams are located 100 feet or greater below the surface. This type of mining requires that miners work underground in mine shafts and extract coal by hand. Deep mining is very dangerous to those who work in the mines. Mine shafts can collapse, trapping and killing laborers. Mine shaft explosions can also occur from volatile gas build-up and poor ventilation. Despite numerous safety regulations, this is still a frequent occurrence around the world. Indeed, in countries that lack adequate safeguards, a great number of miners are killed in this manner. China's coal industry is the most hazardous in the world, officially registering 6,000 deaths in 2004 (Watts 2005). In the same year, the United States registered some 28 deaths in the coal mining industry (MSHA 2006). Another hazard posed to miners is risk of black lung disease. This respiratory illness is caused from chronic (i.e., long-term) exposure to mine dust. Although it is not always fatal, black lung disease reduces life expectancy and makes miners more susceptible to respiratory illnesses, including emphysema, bronchitis, pneumonia, and tuberculosis (Cohen 1990, 36).

In addition to the occupational hazards of deep shaft mining, there are environmental problems. Land subsidence, where land sinks down or collapses into abandoned mine shafts, has been shown to occur in areas with numerous shafts. This settling can damage structures and be dangerous to humans if abandoned mines are not marked. It is estimated that approximately 2 million acres above coal mines have subsided (Cohen 1990, 35). Abandoned mines pose pollution problems, too. As water seeps into abandoned shafts, it mixes with the sulfur compounds that are in mining residues, creating acid mine drainage. If the mixture leaches into rivers and streams, it can be very damaging to fish and other aquatic organisms. Waste from the processing of coal also can be hazardous. Coal is often washed after it is removed from the ground. This process results in a black residue that can be toxic. Generally this waste is piled near mining sites or coal processing sites for later disposal, but removing these piles is costly. Eventually the piles create a hazard by leaching toxic chemicals into the ground.

In areas where coal is located closer to the surface, the land is stripped away to reach the coal beds below. While surface mining is safer for miners, it is devastating to the landscape. In addition to removing all vegetation on the land, the top layers of soil (those with the nutrients available for plant growth) are removed. After mining operations are complete, regeneration of the landscape takes many decades. Restoration projects have been implemented in many areas, but many thousands of acres of land remain scarred from strip-mining operations.

Strip mining is especially damaging in the process of mountaintop mining and valley fill operations (MTM/VF), a practice that is widely used in the Appalachian region of the United States. Mountaintop mining is a surface-mining procedure that strips large portions of land off of mountaintops to reach low-sulfur coal veins found below. Valley fill is the rock and debris, or excess spoil, removed from the surface that is often difficult and costly to return to the mountain and so is instead dumped in adjoining valleys. The valley fill procedure has been criticized for its adverse effects on headwater streams. A draft Environmental Impact Statement (EIS) released by several U.S. federal and state agencies estimated that mountaintop removal has directly impacted 1,200 miles of headwater streams and that biological assemblages of fish and invertebrates are often less diverse in watersheds impacted by MTM/VF (USEPA 2003).

Oil and natural gas are often (but not always) found in the same areas; hence development of oil fields often occurs congruently with the drilling for natural gas resources. This process can be disruptive to natural environments. The heavy equipment and infrastructure needed to drill and pump oil create erosion problems for fragile ecosystems. Additionally, much of the development of these resources is done offshore along continental shelf regions in the oceans. These areas are very biologically productive, and fishing industries depend on their vitality for harvest.

Offshore drilling involves initial surveying and exploration, development and production of the resource, and final decommissioning of the drilling rig when the well no longer produces oil. All of these stages can be damaging to marine life. Exploration techniques utilize seismic and electrosurveys, which impart physiological damage to marine organisms. Well development involves drilling operations that pollute the surrounding environment. Although some care is taken to minimize pollution, it is estimated that annual worldwide loss of oil at sea during stages of extraction is 7 million tons (Patin 1999, 35). Pollution of hydrocarbon compounds, drilling muds (industrial fluids that lubricate the drilling process), and cleaning fluids can be toxic to marine organisms and hence pose a threat to the fishing industry. When a well does not produce any more oil, it must be capped and the infrastructure removed from the offshore area. Unfortunately, this is an expensive endeavor, and many times the equipment remains in place for long periods.

Pipelines and oil tankers are the two main modes of transport for petroleum. Oil spills are one of the most damaging and highly publicized environmental issues associated with oil. Spills occur both on land and in the ocean and are caused by accidents on oil rigs, grounding of oil tankers, and intentional attacks against oil pipelines. It is estimated that over 45 million tons of oil per year are lost to the environment during the various stages of production and transportation (Patin 1999, 35). While significant losses occur in the ocean and on land, most of this oil (22 million tons) is lost to land spills. Because of the danger of spills, many people oppose the construction of oil pipelines across pristine wilderness areas. For example, in the 1970s, several environmental groups opposed the construction of the Alaskan Oil Pipeline because of the potential damage that a spill could cause to pristine wilderness areas. The oil crisis of the 1970s eventually allowed for the construction of the pipeline, but the debate over Alaska's wilderness area remains. The debate is discussed in greater detail in chapter 3.

Oceanic oil spills are very damaging to coastlines and marine environments. The history of these spills dates back to when companies first began to extract and transport oil, but major oil spills began receiving a greater amount of press coverage in the second half of the twentieth century. Oil spills can occur from blowout accidents on tankers and oil rigs. (A blowout occurs when a sudden and uncontrollable discharge of oil or gas erupts from a well or offshore drilling platform.) Such was the case off the coast of Santa Barbara in 1969 when a blowout on a well platform released 230,000 gallons of crude oil, polluting Southern California beaches (Rothman 1998, 101). The Santa Barbara spill pales in comparison to a blowout that occurred from the IXTOC 1 well in the Gulf of Mexico in 1979. This incident released 140 million gallons of oil into the sea, the largest amount ever recorded from an accidental release (Gorman 2001, 330).

Human-related accidents have also caused large oil spills. The most publicized of these is the grounding of the *Exxon Valdez* oil tanker in Prince William Sound in Alaska. The accident caused some 11 million gallons of crude oil to spill out over 900 square miles into the ocean (Smith 2004, 156). The sound, which was a vital breeding ground for migratory birds, became a precarious and dangerous habitat for the waterfowl. Besides being toxic, the crude coated animals in black residue. The oil suffocated fish by clogging their gills and coated the feathers of birds causing them to drown. Because of the residue, many marine mammals perished from the oil spill and the local fishing industry was devastated. Other human-related oil spills have released larger volumes of oil. The 1967 accident with *Torrey Canyon* in the English Channel (36 million gallons) and the 1978 breakup of the *Amoco Cadiz* off the coast of France (65 million gallons) were both larger than the *Valdez* spill (Gorman 2001, 335). But because it occurred in such a biologically diverse area after oil companies had sworn to adopt higher safety standards, the *Valdez* spill was a greater devastation in the eyes of the public. It demonstrated that the extreme negative effects of oil pollution and the difficulties inherent in preventing oil spills had yet to be resolved.

As the discussion above demonstrates, many environmental and human health effects arise from the production and transportation of fossil energies. The brief outline only scratches the surface of the intricacies of these problems. It is important to remember that all of the processes and operations involved in the extraction and transportation of coal, oil, and natural gas also require energy inputs. This energy comes from the combustion of fossil fuels, which produces air pollution.

Air Pollution and Atmospheric Deposition

The combustion of fossil fuels is not only necessary to provide energy for the extraction process, it also is the main way that stored chemical energy is converted from these fuels into useable forms, such as electricity and transportation. Air pollution was one of the first recognized problems with the burning of fossil fuels. Black smoke poisoned the air surrounding early industrial cities. It caused concern among citizens of urban areas. Beginning in the late 1800s, antipollution coalitions worked to raise awareness to public officials. These movements stimulated the first air pollution control laws in Europe and North America. During the latter half of the twentieth century, many countries implemented controls on the pollution emitted when fossil fuels are burned. However, there is still a long way to go. What are the pollutants that are released from fossil fuel combustion? What harmful effects do they have on our environment and human health? This section answers these questions.

The burning of fossil fuels releases a variety of harmful pollutants into the atmosphere. The most common emissions are carbon monoxide (CO), sulfur dioxide (SO_2), nitrgen oxides (NO_x), ozone (O_3), and particulate matter. In addition, toxic metals (such as mercury and lead), hydrocarbons, and volatile organic compounds (VOCs) are released into the atmosphere. Many of these chemicals have been shown to have severely negative effects on human health. Others create hazy smog, which reduces visibility in many areas. Table 2.1 summarizes the different categories of air pollutants.

Different types of fossil fuels release different pollutants when combusted. Coal is the most polluting fuel; however, because it varies in chemical composition depending on the region it is from, the amount of pollution produced varies. Coal combustion produces large quantities of fly ash that most commonly contains CO, particulate matter, SO_2, and NO_x. In modern power plants much of this ash is collected before it is emitted from smokestacks. Since it is impossible to remove all of the pollutants from an emissions stream, regulations still allow for a permissible level of pollution to be emitted in a particular region. These regulations have reduced pollution significantly, but it is important to point out that not all countries have the same regulations. Recall the case of the former Soviet Union, whose archaic industries were not only inefficient, they were also highly polluting.

TABLE 2.1
Common Air Pollutants and Their Environmental and Health Effects

Pollutant	Impact on environment and human health
Sulfur dioxide (SO_2)	Associated with and shown to exacerbate respiratory illnesses (asthma, emphysema, bronchitis, etc.); causes acid rain when it reacts with water in the atmosphere
Nitrogen oxides (NO_x)	Irritates lungs, causes bronchitis and pneumonia, elevates pulmonary edema levels, and lowers resistance to other respiratory infections. Contributes to smog formation
Carbon monoxide (CO)	Affects the body's ability to assimilate oxygen; increases the risks for heart disease and impacts brain functioning
Particulate matter	Scratches/damages the respiratory system; can cause acute or chronic respiratory illness. Some particulates (e.g., benzo[a]pyrene) can cause cancer
Hydrocarbons	Causes smog and contributes to the formation of ground-level ozone
Ozone	Irritates the eyes and mucous membranes of respiratory tract; damages immune system; causes pulmonary congestion, chest pains, coughing; can react with NO_x to form smog.
Volatile organic compounds (VOCs)	Inhalation may cause cancer; can react with NO_x compounds to form smog
Toxic metals (Cd, Pb, Hg, Ar, Ni, Cr, etc.)	Most cause neurological damage to humans and are especially harmful to pregnant women, neurological development of fetuses (increased instances of mental retardation, etc.), and children; can contribute to high blood pressure, heart disease, respiratory illness.

Carbon monoxide is the most common pollutant that is formed when fossil fuels combust completely. In high concentrations, it can impair the ability to function properly, causing drowsiness and headaches. Most people are never exposed to this level of CO, but chronic exposure over time has been shown to increase the rate of heart disease.

Particulate matter (PM) is classified according to size. For example, PM-10 and PM-2.5 refer to particulate matter that is 10 and 2.5 micrometers in diameter, respectively (smaller than the period at the end of this sentence). Scientists have discovered that the smaller the particle size, the more easily it can become embedded in lung tissues of humans. PM can cause damage to the tissues of the respiratory system and is known to cause or exacerbate many respiratory infections. Sulfur and nitrogen oxides can also cause respiratory problems. Because of smokestack and vehicle emissions, urban air pollution can cause life-threatening health problems. Studies have shown that between 30,000 and 35,000 Americans die prematurely because of air pollution (Smith 2004, 84).

Smog is another problem that occurs with the release of these pollutants in the atmosphere. It forms when nitrous oxide molecules (NO_2, N_2O, etc.) react with ozone and water vapor in the atmosphere. Smog produces a brown haze that settles over areas where NO_x is being emitted. It can cause hazy days and impact visibility. VOCs can also react with NO_x to form ground-level ozone. Although ozone is a vital part of the upper regions of Earth's atmosphere, at the ground level it damages the mucous membranes of humans, creates respiratory problems, and can irritate the eyes.

Air pollution can also cause significant damage to crops. Crop yields can be reduced by as much as 10 to 15 percent in polluted areas. Ground-level ozone is the most damaging of all air pollutants to agricultural yields. For example, in the southeastern United States, where O_3 concentrations can reach between 50 and 55 parts per billion (ppb), 10 percent reductions in cotton, soybean, and peanut crops were observed (Smith 2004, 86). In Spain, watermelon yields decreased by 19 percent when O_3 concentrations were measured above regulatory limits.

Atmospheric deposition is one of the most pervasive and damaging consequences of using coal for the generation of electricity. Many bituminous coal resources contain high amounts of sulfur. When this sulfur is burned, it creates sulfur dioxide (SO_2) gas, which is emitted with the exhaust from the combustion. Acid rain occurs when SO_2 reacts with water vapor in the atmosphere to form sulfuric acid (H_2SO_4). During storm events, acid rain is deposited into water bodies and on land surfaces. It is damaging to ecosystems and plant life and can cause increased rates of erosion in some areas (Alexander 1996, 21).

Deposition of solid forms of pollution can also be damaging to ecosystems. Petroleum and coal contain many impurities, including trace metals (for example, lead and mercury) that are toxic to humans. This toxicity is the reason why unleaded gasoline fuels vehicles. But while lead concentrations have been regulated in petroleum products, mercury emissions are an increasingly important problem for countries that rely on coal-fired power plants to supply their electricity needs. Once mercury is emitted into the atmosphere, it can be transported long or short distances, depending on what chemical form it has assumed. Mercury is eventually deposited onto land or water surfaces, where it can settle into sediments and be chemically transformed by microorganisms. This process is called methylation (USEPA 1997, vol. 3, 2–5). Methylated mercury bioaccumulates in the tissues of organisms and becomes more concentrated as it moves higher up the food chain. Studies reveal that methylated mercury is extremely toxic for humans. It is considered to be a potent neurotoxin (it damages the nervous system) capable of causing developmental problems in human fetuses and cognition problems in adults. In ecosystems, methylated mercury can cause developmental damage to several species of migratory birds and can impair the developmental biology of fish (USEPA 1997, vols. 5–7). The pervasiveness of mercury is the reason many governments issue fish-eating advisories to their citizens.

The above example illustrates the complex nature of how air pollutants behave in the environment. There are many other pollutants emitted during the combustion of fossil fuels than those mentioned. These pollutants, which can damage humans and the environment, include other metals toxic to humans and volatile organic compounds, many of which are known to cause cancer. Discussion now turns to global climate change, which is one of the most pressing global environmental problems caused by fossil fuel combustion.

Global Warming

The problem of global warming receives a lot of publicity. Movies like *The Day After Tomorrow* released in 2004 portray outrageous events that can happen if measures to mitigate global warming are not taken. While this movie may be sensational and unrealistic, it does highlight an important and sobering fact: reliance on fossil fuels is altering the global climate, and the consequences of this change are potentially damaging to humans and the Earth's ecosystems.

The phenomenon of global warming is defined as the warming of the Earth's surface temperature by several degrees. This warming can have serious implications for weather patterns and has been shown by many studies to impact ecosystem health and biodiversity. It is a problem directly linked to energy consumption from fossil fuels.

When fossil fuels are combusted, they release carbon dioxide (CO_2), methane (CH_4), water (H_2O), and chlorofluorocarbons (CFCs) into the atmosphere. These gases are often called greenhouse gases because they absorb infrared radiation from the Sun and hence do not allow it to escape (Alexander 1996, 20). Because heat that is normally reflected is trapped in molecules of these gases, a warming effect occurs on the Earth's surface. A great amount of evidence demonstrates that global warming is occurring. The International Panel on Climate Change (IPCC) has determined that surface temperatures on Earth have increased approximately 0.6 degrees Celsius (approximately 1.1 degrees Fahrenheit) since 1861, mountain glaciers have retreated, and sea levels have risen between 0.1 and 0.2 meters in the twentieth century. Furthermore, the IPCC estimates that these changes have occurred more rapidly than any other warming event during the last 1,000 years. This change has been largely attributed to the industrialization of society. The burning of coal, oil, and natural gas releases substantial amounts of CO_2 and CH_4 into the atmosphere. In the last 150 years, atmospheric CO_2 concentrations have increased from 280 parts per million (ppm) to 350 ppm, and most of this increase has occurred since 1960 (IPCC 2001). Additionally, CH_4 concentrations have increased from 0.8 ppm to 1.7 ppm in the past 150 years. While concentrations of CH_4 are lower than those of CO_2, methane is more effective at retaining heat. Most scientists now agree that global warming is a serious issue that needs to be addressed in energy policies (UCS 2005).

Global climate change has the potential to significantly alter the way humans live on the earth. Although these effects are widely debated in the scientific community, concerns include an increase in frequency and intensity of severe weather events (such as hurricanes), an increase in drought cycles creating stress in regions that already have a limited supply of freshwater, and a shifting of many of the world's ecosystems resulting in a loss of biodiversity. Rising sea levels will inundate coastal cities (where a large majority of the world's population resides), a problem that will create an incredible economic burden. Finally, many models of climate change predict that disease outbreaks will become more frequent and severe.

While it is difficult to determine the severity of these impacts, there is a large amount of scientific evidence that supports these predictions. Because of these potential consequences, many governments are looking for energy resources that reduce greenhouse gas emissions. Nuclear energy has been proposed as an alternative, but it too has a variety of adverse environmental and health impacts.

Nuclear Energy

Nuclear energy has often been proposed as the ideal alternative to fossil fuels. It does not produce harmful air pollutants or greenhouse gases, and many energy security issues (described in the next section) would be resolved with the introduction of more nuclear power sources. However, despite the positive attributes of the technology, there are other environmental and health issues associated with nuclear energy. This section examines these issues, which include the harmful effects of radiation, the dangers of a nuclear power plant explosion, and the problems that arise with disposal of nuclear waste.

Radioactivity and *radiation* are words that many people have heard before but have little understanding of their meaning. Chapter 1 describes how a nuclear reaction is able to take place because of the radioactive properties of certain elements like uranium and plutonium. These elements spontaneously emit energy in the form of particles, or rays. In other words, the reaction happens without any stimulus. Although the term radiation is used to describe all types of energy (e.g., light energy or microwave energy), it is ionizing radiation that is of concern with nuclear fuel and weapons. Radiation emitted from radioactive elements can be dangerous for humans if they are exposed to high levels of radiation. Such exposure is known to drastically increase a person's risk for developing cancer. Some types of radiation also have adverse effects on human and animal cells. They cause mutations to occur in the structure of DNA (deoxyribonucleic acid, the basic building block of all life), which may result in genetic effects that can be passed on to future generations. If humans are exposed to extremely high levels of radiation, they may develop radiation sickness. This illness results in the loss of function from major bodily organs. If too many cells in an organ are exposed to too much radiation at one period of time, the organ will die. It is important to note that a person's risk of being exposed to this much ionizing radiation in their lifetime is extremely low. These effects are mentioned here so that the reader understands why high levels of radiation exposure are dangerous and why there are public concerns over radiation levels in the environment.

Accidents at nuclear power plants are an extreme illustration of the problem with nuclear power generation. Accidents are dangerous because they could release high levels of radiation into the environment. Because of their severity, the development of nuclear power has been discouraged in many countries, including the United States. The most common way that a nuclear accident occurs is from overheating problems. If the coolant through a nuclear reactor is not operating properly, then the reactor will overheat. When this happens, the reactor can potentially be damaged, resulting in a number of possible catastrophic events, including explosions that could release large amounts of radioactive material to the environment. The impacts from a nuclear meltdown are widespread. (A meltdown occurs when the core of a nuclear reactor overheats and melts.) If radiation from such an accident becomes airborne, not only will a large number of people be exposed to toxic levels of radiation, but it will contaminate crops and livestock, creating a public health crisis impacting entire populations.

There are only a few instances of nuclear meltdowns in global history. A significant scare occurred in March 1979 at Three Mile Island in Pennsylvania. A faulty valve resulted in a loss of coolant to a reactor. The coolant loss was not discovered for several hours. By the time the malfunction in the reactor was discovered, the core of the reactor had reached 5,000 degrees Fahrenheit and the top of the reactor had melted (Rothman 1998, 146). Plant operators flooded the reactor with water, which immediately turned to steam and destroyed the remainder of the reactor. Radioactive steam was released into the atmosphere, but most of the radioactive material was contained within the reactor. While nobody was killed in the accident, the incident fueled public fear of nuclear power. It was not until 1993 that cleanup crews finished evaporating the radioactive coolant and the reactor was sealed (Manahan 2000, 575).

A far worse accident occurred in 1986 at Chernobyl in the Russian Ukraine when a nuclear reactor exploded, releasing a plume of radioactivity all over Europe. The explosion, caused by excessive heat and pressure that had built up in the reactor, created a large fire that threatened other reactors at the plant. Emergency workers who rushed to the accident to put out the fire were exposed to deadly levels of radiation. Thirty-one people died as a result of the accident (Cohen 1990, 111). The indirect damage to human health was never quantified, but radiation spread throughout Europe, contaminating crops in regions as far north as Scandinavia (Manahan 2000, 576).

In addition to the threats to public health that can occur from a reactor meltdown, radioactive waste produced from nuclear fuel reactions poses a problem for waste managers. Because spent fuel rods can release lethal amounts of radiation for an extremely long period of time, the waste from these operations needs to be stored in a safe enclosure. Waste that is produced from nuclear reactors can be handled in two ways. It can be sent to a reprocessing plant where fuel can be recycled, or it can be sealed in solid containers, usually made of glass, and stored in underground vaults (Cohen 1990, 177). A number of issues need to be considered in choosing the process for radioactive waste management, including transportation concerns and location of waste repositories. Since waste is often stored underground, the latter concern involves selecting sites that are geologically sound and where there will be minimal risk of groundwater contamination. Controversy over one such site in the United States, Yucca Mountain, is discussed in chapter 3.

Renewable Technologies

As mentioned before, all energy sources impact the environment in one way or another including renewable energy resources. Hydropower is probably the most damaging to the environment. Often large dams are built to capture and store running water. These dams can inundate sizeable areas of land and alter the ecosystems of streams and rivers. While the benefits of dams may outweigh the costs for some people, the drawbacks can be devastating for downstream users. For example, dams along the Colorado River (and others) provide cheap water and power to large cities in the southwestern United States. However, water in the river is overallocated. It only intermittently flows to the Sea of Cortez, and farmers in Mexico receive a fraction of what they once did. By the time the water reaches southern Arizona, it is so high in salinity and other pollutants that a desalinization plant was installed in Yuma, Arizona, to purify the water before it crosses the border into Mexico.

Dams can also pose a problem for fish. Large reservoirs that fill up behind dams present enormous obstacles to migrating fish. Salmon are a good example. Their lifecycle involves a period in which they migrate from the ocean to streams to reproduce. By using chemical cues from their environment, these fish amazingly return to the same stream from which they hatched to spawn a new generation. However, when a river is dammed, the fish need to navigate through countless obstacles, including large reservoirs and high dams, to return to their spawning grounds. Many fish have perished on this journey. Because of dammed rivers, salmon and some other fish species are listed as threatened with extinction. Not only has the impact of dams been an important issue for wildlife conservation groups, it is one of concern to the fishing industry.

Dam issues are also important in many developing countries where large dams can potentially displace millions of people. Protests in India over the Narmada Dam projects halted the construction of several large dams because they threatened to displace thousands of people. Perhaps the most egregious example of displacement is occurring in China, where the construction of the Three Gorges Dam—what will be the largest hydroelectric dam in the world when it is completed in 2009—will displace an estimated 1.9 million people (IRN 2006). The project is so controversial that the World Bank withdrew its financial support (Khagram 2004, 175).

Wind and solar power also have environmental impacts. Wind turbines have been criticized by environmentalists as being a danger to migratory birds. Studies have also shown that large wind farms can impact microclimates because they displace large amounts of air. As for the issues of solar energy, photovoltaic cells pose a problem because the most efficient designs contain toxic chemicals that must be treated properly when disposed. Solar energies are also criticized for the large area of land that a solar power plant needs to produce an economically viable amount of power. This argument, however, is weak when one considers the many millions of rooftops that could generate domestic sources of power in the desert regions of the world. Finally, biomass energy can create problems if it is not managed properly. Historically, there are many instances of humans depleting their biomass resources. Depletion remains a problem in many developing countries where wood-gathering provides a main energy source used for heating and cooking.

Social Problems

In addition to the numerous environmental problems described above, global energy use raises social concerns. Like the environmental issues, the root cause of many of the social problems discussed below lies in the fact that fossil fuels are the primary energy resource used. This section examines these problems. They include diminishing resources of fossil fuels, conflicts that arise due to concerns over energy security, and inequality in worldwide energy availability. It is important to keep in mind that many of the issues discussed are complex and multifaceted. While energy use is not the sole cause of the issues, energy-sector dynamics often work to exacerbate many of these problems.

Diminishing Resources

In order for energy to work for society, it must be affordable, available, and reliable. In general, an energy source is affordable if it is abundant and accessible. Energy is available to a society if that particular society has the proper means of obtaining energy resources, converting them into energy services and products, and distributing them. When the conditions of affordability and availability are met, and as long as the energy systems are maintained, then an energy resource is reliable. Herein lies another problem of fossil fuel dependence: the resource base for these fuels, particularly petroleum, is diminishing. These resources are finite and are being used at a much faster rate than they are being replenished—a situation that will make it harder for countries to meet the three conditions listed above. This discussion focuses on petroleum because it is the main fossil fuel commercially consumed worldwide.

It is harder and harder to find new areas of abundant oil reserves. New discoveries, such as those on the North Slope of Alaska and in the North Sea in the 1970s, often prove to be overexaggerated. Examples of other overstated potential oil deposits include the South China Sea, the Caspian Sea, and Baltimore Canyon (an offshore area located off the eastern coast of the United States) (Smil 2003, 189). Furthermore, as humans search in more remote areas, it is questionable whether the resource can be extracted in an economical manner. As a result, oil may no longer be affordable.

The consequences of diminishing oil resources are debated among economists and energy policy analysts. Pessimists in the debate predict a rapidly approaching "end of oil" and forecast dire consequences if society does not start to address the energy resource problem. They consider the reduction in available oil and increased global demand (especially in light of China's development) as evidence that a major global energy crisis is looming.

Optimists, on the other hand, contend that resources are not so close to being diminished. Estimates are continuously revised to reveal increasing amounts of recoverable petroleum. Because techniques for discovering, extracting, and estimating oil reserves are constantly improving, optimists think society will discover more oil to support its needs. Even when oil resources are no longer recoverable, human ingenuity will discover ways to supply necessary energy. Instead of drastic consequences, society will slowly adapt to different energy resources.

This debate is important because it demonstrates the two main sides of the argument involving oil resources. Regardless of the analyses done by both sides, oil resources are finite and at some point they are going to run out. In 1996, it was concluded there are only 850 Gb (gigabarrels) of remaining oil, which is between 17 and 27 percent less than was originally thought at the time (Smil 2003, 191).

What fossil fuel will replace oil? Perhaps natural gas, which is considerably more abundant than oil, will. Natural gas burns cleaner than either coal or oil, and many economists predict that this resource will play a much more important role in the future. On the other hand, coal's future is not so bright. Despite its abundance, concerns over pollution and global warming have caused many countries to develop other energy sources, mainly nuclear.

There is no question that the Earth will eventually run out of oil. In the meantime, society needs to be concerned with issues related to an increasingly scarce commodity. Conflict and energy security issues are the next social problem discussed.

Security Issues and Conflict

The movie *Syriana*, released in 2005, told the deeply political story of how the United States relates to petroleum-producing countries. In the movie, the U.S. Central Intelligence Agency is working to secure the favor of a fictional Middle Eastern country. Tension arises when a new member of the royal family of the country is chosen to ascend the throne. The movie unveils the corruption involved in the oil industry and how oil deals between countries can be tied to terrorist activities. Although the movie was fictional, it illustrates the importance of energy security issues.

Conflict between countries is historically linked to resource acquisition. Chapter 1 reveals that many of the energy crises in the latter half of the twentieth century were caused from military conflicts. First, the 1956 Suez Crisis brought about an energy emergency in Europe. Then, the Arab Oil Embargo in 1973 caused the worst energy crisis in U.S. history. While energy resources were not the cause of these conflicts, energy supply was greatly impacted, raising alarm over the extent of U.S. reliance on imported fuels. Not only are industrialized countries already dependent on foreign sources of oil, petroleum imports to these countries are expected to increase by 70 percent by 2020 if current demand remains the same (Gellar 2003, 11). These numbers raise alarm with many energy security analysts when they consider the tensions that exist in many oil-producing countries. The United States imposes economic sanctions against many of these countries for various human rights abuses and terrorist activities (Smith 2004, 148). The question inevitably arises, if these nations are so unstable, how wise is it to rely on them for energy needs?

Energy security is of primary concern for industrialized countries, and military intervention is being used more and more to secure oil supply. Because energy availability is a vital part to economies, industrialized countries will go to great lengths to ensure their supply. Most notable in recent history is the Desert Shield/Desert Storm operations launched by the United States in 1991–1992. The conflict began when Iraq invaded the small country of neighboring Kuwait. If it had succeeded, Iraq would have doubled its petroleum reserves (Smil 2003, 119). Such an acquisition would have not only threatened oil supply to the United States, it would have challenged Saudi Arabia's (a key U.S. ally) hold on the oil markets. The United States responded to the Iraq invasion of Kuwait by launching two coordinated offenses against the Iraqi government, Operation Desert Shield (armed deployment) and Operation Desert Storm (bombing campaign and ground offensive) (94). While it is difficult to pinpoint a reason for any military conflict, analysts concluded that energy security played a large role in the U.S. military intervention in Iraq in the early 1990s.

The problem of nuclear proliferation also raises energy security concerns. Because of its nonpolluting attributes and cheap operating costs, many developing countries have expressed interest in developing their nuclear energy capabilities. However, in light of increased terrorist attacks, concerns over the intentions of some countries have surfaced. The possibility that these nations will develop nuclear weapons in conjunction with nuclear fuels raises fears in industrialized countries. While the United Nations works to provide peaceful resolutions to these variances, it seems apparent that disagreements over these issues are heightening. Countries seeking to produce nuclear power are becoming increasingly frustrated at attempts to halt their development. At the same time, industrialized countries are becoming more convinced that illicit activities are occurring in other countries. For example, the United States invaded Iraq in 2003 using the rationale that Saddam Hussein's regime was harboring weapons of mass destruction. When evidence surfaced proving this claim false, many critics contended that oil security was the reason for the invasion. Either way, this example is another demonstration of the conflicts that result over energy security concerns.

Inequalities

The issue of energy availability is important not only in the sense of what energy resources can be extracted but also in who gets these resources and how does that impact those who cannot afford them. These concerns arise because of the increasingly inequitable trend in energy use. Rich, industrialized countries have all of the energy they need and then some. They have cheap gasoline supplies and reliable electricity grids. Most people living in industrialized countries are able to afford the energy they need for heat, lighting, and cooking. If they are sick, they can go to well-lighted and clean hospitals. The roads they travel on are paved, and they have access to many different modes of transportation. Energy is so abundant in the United States that its citizens often take it for granted, not fully understanding the extent of energy poverty in which others live.

It is estimated that over 2 billion people worldwide cannot access basic energy services, relying instead on biomass energies to support their needs (Goldemberg and Johansson 2004). The necessity to find energy sources is most cumbersome to women and children who spend a large part of their day gathering wood and water. Not only is this physically burdensome, it limits their educational opportunities as a large part of their time is devoted to just trying to survive. Because their countries do not have established energy infrastructure that provides transportation, these people are rarely able to travel. Lack of access to energy services also impacts the quality of health care. For these reasons, many people lack the opportunities and abilities to change their economic status. It is in this way that energy inequality exacerbates poverty. Furthermore, the income gap between rich and poor countries is increasing due to energy inequality. If poor countries are not able to develop their energy infrastructure, their poverty will only increase.

The energy inequality problem is also tied to the trend of globalization discussed in the first section of this chapter. While globalization may help countries develop their energy resources, the increasingly scarce quantities of fossil fuels and the unstable nature of energy prices could be detrimental to those countries seeking to participate in global energy markets. Because of the potential for resource-related conflicts, energy prices are predicted to become more volatile in the future. Reliance on these markets to provide energy services could be detrimental to countries that are trying to develop their infrastructure. Recall that, in general, the initial stages of economic development require greater energy inputs (figure 2.1). If these inputs are unreliable because of unstable prices, then developing countries could be in a precarious position and may fall even more behind in their development prospects, all the while increasing their foreign debt and further impoverishing their people.

Concerns of inequality also embody the means by which energy resources are extracted. As discussed in the energy and economics section, the globalization of the economy has allowed many energy companies to establish energy extractive resources in developing countries. Many poor countries do not have the same safety and environmental regulations that richer countries have. Because of inadequate environmental and safety standards, energy extraction industries can pose a great threat to the citizens of poorer countries. While the energy resources extracted by TNCs are exported to richer countries to supply their needs, the citizens of poorer countries rarely see improvements in their own energy infrastructure. They do see pollution increasing human health problems and degrading arable land. They must endure the negative consequences of energy resource extraction without receiving the benefits of energy services.

It is difficult to understand the depth and intensity of inequality issues that relate to global energy use. Energy inequity is related to global economic phenomena and is also intertwined with how energy use affects the environment. Despite this complexity, there are solutions that can be implemented by society to alleviate the negative effects of energy trends. These are examined in the next section.

Solutions

The previous section illustrates the necessity to change the way society consumes energy. It is apparent that social and environmental problems caused by energy consumption are complex and that there is no one solution to alleviate them. However, society can take action to mitigate negative impacts. These solutions range from dynamic, long-term shifts in the way that society operates, to the implementation of national energy and environmental policies, to the personal choices that you and I make on a daily basis. This section discusses the various options for improving the outlook of energy use in the future. First is an examination of the concept of sustainable development, a notion that requires dramatic changes in the way society thinks about energy resources. Second is a discussion of the necessity for a transition to a renewable energy resource base. Third is a discussion of policy options at the national and global levels. The chapter concludes with an examination of the personal changes that people can make in their lives. It is important to note that all of these options (as opposed to one or another) must be considered in the design of future energy solutions.

Sustainable Development

The idea of sustainable development challenges the dominant way in which humans view progress and economic growth in society. This paradigm asserts that society not only needs to change the way that it approaches energy, it also needs to examine the foundations of economic progress that underlie the way people use resources. This approach requires a shift in the ways that energy resources are developed, distributed, and utilized. It requires that energy use be organized in such a way that it does not affect the ability of others to use energy and does not jeopardize human health or the environment. There are both ideal and practical ways of thinking about the concept of sustainable development.

Sustainable development as an ideal embodies an ethic that respects the environment and human rights. It contends that all people living on this planet (and future generations to come) should have the same opportunities for enhancing their quality of life, and that in pursuing opportunities, they do not affect the ability of others to do the same. This goal cannot be met with mainstream, unsustainable models of economic growth. These models are based on an anthropocentric philosophy that views natural resources as tools for human progress. Because of this view, environmental and social costs of energy production and consumption are not considered in projections of future growth. Unsustainable models of economic development do not value environmental and human resources; they violate human rights and diminish environmental quality.

Conversely, sustainable models for economic growth consider that healthy environments provide valuable services to humanity. Environmental health must be a factor in economic analyses because clean air, clean water, and productive lands provide long-term benefits for humans. Because of its inclusion of environmental factors, the philosophy of sustainable development promotes a way of approaching energy use that is fundamentally different than that of common economic growth patterns. It promotes conservation of resources and the development of alternative, eco-friendly energy services. Additionally, asserting that access to energy services is a basic human right, sustainability philosophies seek to reduce poverty and alleviate human suffering from poor energy practices.

Practical ways of using the concept of sustainable development involve increasing energy efficiencies and relying more on renewable energy to supply the needs of society (Gellar 2003, 16). Furthermore, sustainable approaches recognize that a variety of energy resources can be used by societies. Rather than adopting conventional, commercial techniques for energy provision (i.e., oil, coal, etc.), sustainable approaches empower local communities to develop and maintain energy sources that best fit their society. Practical sustainable development also promotes conservation. It requires that people and countries change wasteful energy habits. These approaches seek to raise awareness about energy-saving lifestyles and help people obtain sustainable development goals.

The concept of sustainable development has spilled over into approaches for a society to take to alleviate energy concerns. The next sections discuss the shift toward utilizing renewable energy sources, how sustainable ideas can influence energy policies and global norms, and how sustainable philosophies have motivated personal choices.

Technology Solutions: Transition to Renewable Sources

The discussion in chapter 1 on the history of energy use in part focused on the different energy transitions that have taken place throughout history. It discussed historical trends in energy use as being both the cause and the result of major energy transitions. Recall that the term *transitions* refers to a shift in the primary resource that a society uses to obtain energy. Energy transitions are long-term changes in the way that society operates. They involve the construction of new infrastructure to harness energy from different sources, the application of new technologies to efficiently utilize that energy source, and the continuous development of energy systems to accommodate changes in the demographics of populations. Transitions occur slowly, but have always been necessary for society to adapt to changing environments and resource needs.

As the world population increases, human society will undergo further energy transitions. In light of the environmental and social problems associated with global energy use, it seems that humans are at the cusp of the next energy transition. Since finite fossil fuels are being depleted, it is inevitable that society will shift its energy resource base to one that is primarily driven by renewable technologies. The development of renewable energies will not only alleviate energy supply issues, it will also address environmental and social concerns. Even renewable technologies have environmental consequences, but these impacts are much less devastating to ecosystems and human health than those from fossil energies. Additionally, many renewable resources can be locally developed and managed, so the need to rely on foreign sources of energy is lessened. Hence, renewable energies do not carry the burden of requiring military intervention to secure supply.

Early research and development of renewable energy resources began in the United States in the aftermath of the oil crises in the 1970s. The United States promoted renewable development in policies such as President Carter's National Energy Policy (discussed in further detail in chapter 3). Despite early initiatives, interest in developing renewable energy technologies in the United States waned in the 1980s. When Ronald Reagan became president in 1980, he emphasized that the productive capacity of fossil fuels in the United States would solve economic problems and alleviate energy shortages (Smith 2004, 152). Hence, energy policy during the Reagan administration focused on increasing domestic production of fossil fuels and emphasized the ability of the energy market to provide cheap and abundant fossil-fueled energy. Although other U.S. efforts have been made in the area of energy conservation and renewable development (most notably the Comprehensive Energy Plan in 1992 passed by President George H. W. Bush), national promotion of alternative energy resources has stagnated. Individual state governments in the United States offer a notable exception, as thirteen states have implemented Renewables Portfolio Standard (RPS) programs that require a certain percentage of electricity sold to users be derived from renewable sources (Coenraads and de Vos 2004, 59).

Despite the U.S. reluctance to support renewable energy, other countries have made significant progress in this area. The European Union (EU) has emerged as the world leader in renewable energy, obtaining approximately 390 TWh (terawatt hours) of electricity from renewable sources in 2002 (Coenraads and de Vos 2004, 58). In 1997, the European Commission issued its "White Paper for a Community Strategy," which outlined a policy of doubling the use of renewable energy (from 6 percent to 12 percent) for electricity generation by the year 2010 (EU 2005). The European Union (EU) expanded this goal in 2001 with the Renewable Electricity Directive, which set the 2010 target at 21 percent. In addition to the EU's multinational directive, fifteen of the twenty-five member-nations have developed and implemented policies that incorporate renewable technology into their electric and transportation sectors.

Two other industrialized nations have also recently endorsed the "green" energy market. In Australia, the Mandatory Renewable Energy Target (MRET) sets a target of 9,500 GWh by 2010, with penalties issued to power companies that do not comply with the program (Coenraads and de Vos 2004, 59). The voluntary Green Power Program in Australia generated 400 GWh of renewable electricity in 2004. In 2003, Japan developed a Renewable Portfolio Standard (RPS) that sets an electricity production goal using renewable energy at 12.2 TWh by 2010.

The efforts to generate renewable electricity are increasingly significant and are expected to be enhanced in future energy pursuits. The development and introduction of new technologies are not the only solutions to society's energy problems. In order to utilize technologies effectively, they must be employed in conjunction with sustainable energy and environmental policies. The next section examines policy issues and global initiatives to promote changes in the energy sector.

Policy Solutions

Energy policies are important at both the national and international levels. Good policies are dynamic and adaptive. They consider issues of implementation and understand that economic and environmental policies are also important in the effectiveness of energy utilization. This section addresses concerns that should be raised for national energy agendas. It then examines global initiatives and international environmental regimes. Finally, it briefly describes the advances that have been made in addressing global climate change.

National Policies

In order to address energy concerns effectively, countries must prepare energy policy in conjunction with environmental and economic policies. Energy use has important implications for economic growth and is closely linked to environmental degradation. Energy policy that does not include economic and environmental considerations will be ineffective in alleviating the problems associated with energy use. Because it is necessary to move away from fossil energies, renewable energy technologies should be promoted as much as possible in energy policy designs.

In constructing national energy policies, countries should consider how renewable energy can be developed using the resources found within a region's own borders. For example, in the United States, solar energy is a good source of electricity for the desert regions of the Southwest. However, this type of resource would not be as effective in the Northwest where cloudy, rainy days are more common. Similarly, tidal power and wave energy technologies can be important resources for coastal cities but are not feasible for inland areas. These examples illustrate that different renewable energy sources can be utilized effectively in different regions. Often this utilization involves decentralizing energy resources and distribution centers to provide greater management flexibility in local regions.

One goal of energy strategy as it relates to economic policy should be to move away from dependence on fossil fuels by increasing the costs of their use at the same time as increasing the accessibility, availability, and affordability of renewable energies. Policies must carefully consider how energy subsidies are distributed. Energy subsidies are most often awarded to fossil fuel technologies. Not only does this promote further development of these resources, it also has the potential to encourage wasteful energy consumption. When the cost of fossil fuel production is offset by subsidy, there is no incentive for either producers or consumers to conserve energy. Alternatively, energy policies that subsidize research and development of renewable resources while at the same time impose taxes on fossil fuel consumption would promote consumption of renewable resources and discourage the use of nonrenewable energy. Carbon taxes, those that impose a fee for CO_2 emissions, are another option that can be used to increase energy efficiency and shifts to cleaner sources. It is important to note that these policy changes need to be phased in over time. Recall that energy subsidies are developed for the purpose of promoting development opportunities to all members of society. If subsidy policies are not implemented properly, they may adversely impact poor people by increasing the costs of energy services.

Environmental regulations are also an important part of a national energy policy because, while their immediate goal is to reduce pollution, they incorporate the environmental costs of fossil energies. It is expensive for companies to install pollution control devices in their energy systems. As a result, environmental protections are reflected in the cost of energy services. Over time, as pollution control technology becomes standard, energy prices reflect the environmental costs of production. So through environmental regulation, renewable energies, whose environmental costs are much less than those of fossil energies, can be promoted. Goals of energy efficiency could also be met with the enforcement of environmental regulations, as it will cost energy producers (and consumers) more to produce more energy than is necessary.

Because energy markets are becoming increasingly globalized, global environmental norms and standards are increasingly important for ensuring sustainable energy development. These initiatives are discussed in the next section.

Global Policy

Earlier sections mentioned the problem of globalized energy companies operating in countries that do not have established or enforceable environmental standards. This lack of national standards becomes a global problem when the environments that are being polluted from energy industries are considered common pool resources. A common pool resource can be conceptualized using the analogy of a common green, as described by Garrett Hardin in his essay "The Tragedy of the Commons." He explains that individual people sharing a common pasture will seek to maximize their own returns from that pasture by grazing as many sheep on the land as they can afford. Inevitably, this practice will lead to a depletion of resources in the pasture and destruction of the land from overgrazing (Hardin, 1968; Smith 2004, 5). In addition to land, air and water are also considered to be common pool resources. Because they are not owned by anyone in particular, polluters of these resources do not have any short-term incentive to limit their emissions; hence they continue to pollute.

Air pollution, ocean pollution, and global climate change are good examples of common pool pollution problems. As these problems have become increasingly worse, there have been efforts to address them. International environmental regimes have been proposed as one solution for alleviating these problems. An international environmental regime can be defined as "social institutions consisting of agreed upon principles, norms, rules, procedures, and programs that govern the interactions of actors in specific issue areas" (Young and Levy 1999, 1). In terms of worldwide energy actors, regimes seek to develop global environmental standards that will be respected by energy companies operating all over the world. The goal of these regimes is to enhance communication and cooperation among different entities to alleviate the environmental and social problems associated with energy use. International environmental regimes work primarily through global institutions, such as the United Nations, to induce acceptance of environmental standards in the operations and practices of TNCs and the countries that they operate in.

The effort to establish oil pollution control measures for vessels carrying crude oil is an example of an environmental regime. The section on production and transportation of fossil fuels described problems of oil pollution in the world's oceans. Largely, it focused on oil spills. In addition to these disasters, oil pollution problems occur from the practice of intentional discharge from oil tankers. After a ship dispensed of its oil, common practice was to fill the ballast with seawater in order to stabilize the tanker for the journey back to the producing region. Residual oil from the tanker would mix with the seawater, which was released once again to the ocean before the tanker received more oil. It is estimated that for every 100,000-ton vessel, approximately 300 to 500 tons of oil were discharged during each shipment (Mitchell et al. 1999, 34). It is apparent that this practice introduced a large amount of oil pollution to the ocean.

In order to deal with this problem, the International Convention for the Prevention of Pollution from Ships (MARPOL) was developed in 1998. Supported by the United Nations, the convention is an agreement by oil-producing and -consuming countries to adopt standards for oil shipments that oil companies must adhere to. MARPOL created rules that banned oil vessels from discharging oil into oceans. It also required that vessels install pollution reduction equipment. In addition, countries were required to establish enforcement measures that allowed them to monitor the actions of oil companies. MARPOL is an effective international regime because it targets the many different actors involved in the problem. It requires standards for oil shipping companies, the parent countries of these companies, and even the insurance companies that insured oil vessels against damage (Mitchell et al. 1999, 85).

Efforts to address global climate change are also taking place in the arena of international environmental regimes. Although concerns over the impacts of global climate change have been voiced since the 1960s, it was during the 1980s and 1990s that these concerns were addressed by policymakers on a global level. The recession of glaciers and ecosystem shifts were noted by scientists to be consequences of rising levels of greenhouse gases. In response to these concerns, the United Nations Framework Convention on Climate Change (UNFCCC) pact was opened for signatures in Rio de Janeiro in 1992 (UNFCCC 2004a). Countries that signed the pact agreed to find ways to bring their emissions of greenhouse gases under levels measured in 1990. By 1993, the pact had been signed by 166 nations. The UNFCCC created the Intergovernmental Panel on Climate Change (IPCC) for the purpose of gathering and assessing information about the impacts of human-induced global climate change.

The most important global effort that has been developed to address the issue of climate change is the Kyoto Protocol. It was developed in December 1997 during a meeting of the UNFCCC. The goal of the protocol was to establish a legally binding document with defined targets and penalties for countries that exceeded their emissions of greenhouse gases (GCMD 2004). Representatives from over 170 countries participated in the meeting. In order for the treaty to be effective, it stipulated that support was needed from fifty-five parties to the convention, including industrialized countries whose emissions totaled 55 percent of all global emissions in 1990 (UNFCCC 2004b). The treaty became effective in 2005 after Russia agreed to its ratification.

Although the Kyoto Protocol is the first step toward a global commitment to reducing greenhouse gas emissions, it has been hindered by the 2001 decision by the United States not to ratify the treaty. The retreat of the United States created controversy around the world, as U.S. emissions accounted for 36.1 percent, making it the largest producer of greenhouse gases. Some states in the United States have since developed their own regulations for greenhouse gases; however, the United States remains one of the few industrialized countries that has not addressed the problem of climate change on a national level.

While international agreements and environmental regimes are important for implementing change at a global level, they often seem too "big" for the individual person to grasp. They deal with numerous actors operating across the world. Often the issues they are addressing are complex, involving scientific inquiries and technological jargon that is difficult for anyone operating outside the fields to understand. It almost seems that these issues are untouchable to individuals because they are so complex. Is there anything that everyday people can do to help alleviate energy problems? The answer to this question is yes. The next section describes decisions that people can make in their everyday lives to limit their own energy consumption.

Personal Energy Responsibility

As a resident of an industrialized country (and likely of the United States, the most energy-intensive society in the world), there are many ways you can reduce your consumption of energy. These measures begin with conserving energy services. As a child, you were probably taught that turning off lights when you leave a room and turning down the heat (or turning up the air conditioning) while you are away from home are ways to conserve energy. Actions like these are simple; once you are in the practice of doing them, they become second nature. For example, walking or riding your bike instead of driving reduces the amount of fossil fuels you need to burn to go somewhere. It is by simple habits like these that you can reduce your consumption of energy resources. By being mindful of the energy resources you are using, you can modify your life to adopt simple practices that can go a long way in minimizing your own energy consumption and helping society reduce its consumption.

The design of your home and the food you eat afford other ways in which you can conserve energy resources. Depending on the region in which you live, you can exploit nature to make your home more energy efficient. For example, adobe-style housing in the southwestern United States is energy efficient because it remains cooler in the summer (when temperatures can get above 100 degrees Fahrenheit) while trapping heat in the winter. Additionally, other designs, such as awnings on windows, can capture sunlight when it is most needed for heat and provide shade in hotter temperatures. Designs like these are simple. They utilize materials that reduce the amount of energy needed from centralized fossil fuel sources.

Another way in which you can reduce your own personal energy use is by eating out less, eating less meat, and supporting community-based agriculture. Restaurants are very energy intensive. To increase food productivity, their kitchens operate at high temperatures. Fast food restaurants are even more energy intensive because they produce incredible amounts of waste in food packaging. Additionally, meat-based items are featured on most restaurant menus. While the choice to become a vegetarian or a vegan is extreme for many people, it is important to realize that reducing your personal consumption of meat reduces the amount of energy that you use in society. Not only do meat processing plants use large amounts of water and energy, the process of breeding animals requires that large fields of alfalfa be planted. Growing crops like alfalfa has its own energy requirements for pesticide production, water irrigation, and planting, tilling, and harvesting. Finally, choosing local food sources is a way that you can reduce your energy demand. Foods sold in supermarkets have often traveled for several thousand miles before they are available on the market. Alternatively, local agriculture operations have a wide variety of foods without the additional energy requirements of long-distance travel. If one were to trace back all of the energy inputs that are involved in processing and distributing food, it would be surprising to realize the amount of energy that goes into just eating in our society! We often take these things for granted, but by recognizing the many ways in which we use energy, we can also find ways in which we can change our habits. Eating one or two vegetarian meals per week and supporting local agriculture rather than buying food from supermarkets that support agribusiness can reduce energy consumption.

There are countless ways in our lives to reduce energy use. By recognizing them and acting on them, we can individually be solutions to energy problems. By example, we can perhaps collectively change our energy ways.

Conclusion

This chapter examined the many, complex issues that are associated with global energy use. First, it established a conceptual framework linking energy to economics for the purpose of understanding energy problems. Next, it exemplified these concepts by examining how they are relevant to energy trends. It then discussed the many environmental and social problems that arise from society's current fossil-fueled energy path. Finally, it considered different strategies that can be employed for alleviating these problems. In this chapter, the focus was on global energy trends and issues. The next chapter focuses on specific energy problems in the United States.

References

Alexander, G. 1996. "An Overview: The Context of Renewable Energy Technologies. In *Renewable Energy: Power for a Sustainable Future*, edited by Boyle, G., 1–39. Oxford, England: Oxford University Press.

Barboza, D., and A. R. Sorkin. 2005. "Chinese Company Drops Bid to Buy U.S. Oil Concern." *New York Times*, August 3, A1.

Clapp, J., and P. Dauvergne. 2005. *Paths to a Green World: The Political Economy of the Global Environment*. Cambridge, MA: MIT Press.

Coenraads, R., and R. de Vos. 2004. "Europe Set the Pace: An Overview of Australia, Japan, USA and EU Green Power Markets." *reFocus* November/December: 58–59.

Cohen, B. L. 1990. *The Nuclear Energy Option: An Alternative for the 90s*. New York: Plenum Press.

Energy Information Administration (EIA). 2000. *Federal Financial Interventions and Subsidies in Energy Markets 1999: Primary Energy*. SR/OIAF/99–03. Washington, DC: U.S. Department of Energy.

European Union (EU). 2005. "Energy for the Future: Renewable Sources of Energy." http://europa.eu.int/comm/energy/res/index_en.htm (accessed July 21, 2005).

Gellar, H. 2003. *Energy Revolution: Policies for a Sustainable Future*. Washington, DC: Island Press.

Global Change Master Dictionary (GCMD). 2004. "Kyoto Protocol FAQ." http://gcmd.gsfc.nasa.gov/Resources/FAQs/kyoto.html (accessed November 13, 2005).

Goldemberg, J., and T. Johansson, eds. 2004. *World Energy Assessment: Overview 2004 Update*. New York: United Nations Development Programme (UNDP).

Goodman, A., and D. Goodman. 2004. *The Exception to the Rulers: Exposing Oily Politicians, War Profiteers and the Media That Love Them*. New York: Hyperion.

Gorman, H. S. 2001. *Redefining Efficiency: Pollution Concerns, Regulatory Mechanisms, and Technological Change in the U.S. Petroleum Industry*. Akron, OH: University of Akron Press.

Hardin, G. 1968. "The Tragedy of the Commons." *Science* 162 (October–December): 1243–1248.

International Energy Agency (IEA). 2006. Oil Market Report, January 17. http://omrpublic.iea.org/ (accessed January 23, 2006).

International Panel on Climate Change (IPCC). 2001. "Summary for Policymakers." *Climate Change 2001: The Scientific Basis.* http://www.grida.no/climate/ipcc_tar/wg1/005.htm (accessed November 13, 2005).

International Rivers Network (IRN). 2006. "Three Gorges Project." http://www.irn.org/programs/threeg/ (accessed January 26, 2006).

Khagram, S. 2004. *Dams and Development: Transnational Struggles for Water and Power.* Ithaca, NY: Cornell University Press.

Manahan, S. E. 2000. *Environmental Chemistry.* 7th ed. Boca Raton, FL: Lewis Publishers.

Manning, R. A. 2000. *The Asian Energy Factor: Myths and Dilemmas of Energy, Security and the Pacific Future.* New York: Palgrave.

Mine Safety and Health Administration (MSHA). 2006. *MSHA Fatality Statistics.* http://www.msha.gov/stats/charts/chartshome.htm (accessed January 23, 2006).

Mitchell, R., M. L. McConnell, A. Roginko, and A. Barrett. 1999. "International Vessel-Source Pollution." In *The Effectiveness of International Environmental Regimes,* edited by O. R. Young. Cambridge, MA: MIT Press.

Patin, S. 1999. *Environmental Impact of the Offshore Oil and Gas Industry.* East Northport, NY: EcoMonitor Publishing.

Rothman, H. K. 1998. *The Greening of a Nation? Environmentalism in the United States since 1945.* Fort Worth, TX: Harcourt Brace College Publishers.

Smil, V. 1998. *Energy in China's Modernization: Advances and Limitations.* Armonk, NY: M. E. Sharpe.

Smil, V. 2003. *Energy at the Crossroads: Global Perspectives and Uncertainties.* Cambridge, MA: MIT Press.

Smith, Z. A. 2004. *The Environmental Policy Paradox.* 4th ed. Upper Saddle River, NJ: Prentice Hall.

Union of Concerned Scientists (UCS). 2005. "Global Warming Overview." http://www.ucsusa.org/global_environment/global_warming/index.cfm (accessed June 21, 2005).

United Nations Framework Convention on Climate Change (UNFCCC). "The Convention and Kyoto Protocol." 2004a. http://unfccc.int/resource/convkp.html (accessed October 21, 2005).

United Nations Framework Convention on Climate Change (UNFCCC). 2004b. "Russia Decision on Ratification: Major Step Toward

Entry into Force of Kyoto Protocol." http://unfccc.int/files/press/releases/application/pdf/pr040930.pdf (accessed November 13, 2005)

U.S. Environmental Protection Agency (USEPA). 1997. *Mercury Study Report to Congress.* Vols. 1–8. EPA-492/R-97–006. December. Washington DC: Author.

U.S. Environmental Protection Agency (USEPA). 2003. Mountaintop Mining/Valley Fill Draft EIS, Executive Summary. http://www.epa.gov/region3/mtntop/pdf/Executive%20Summary.pdf (accessed November 13, 2005)

Watts, S. 2005. "A Coal Dependent Future?" *BBC News,* March 9.

World Bank. 2005. *World Development Indicators Database.* http://www.worldbank.org/ (accessed February 10, 2006).

Young, O. R., and M. A. Levy. 1999. "The Effectiveness of International Environmental Regimes." In *The Effectiveness of International Environmental Regimes,* edited by O. R. Young. Cambridge, MA: MIT Press.

3

Special U.S. Issues

Introduction

Chapter 2 demonstrated how economic activity is linked to energy consumption. It described that in general if a country's GDP (gross domestic product) is high, the country uses a lot of energy relative to other countries. While not always the case, this correlation is accurate for the United States. Globally, despite making up only 4.6 percent of the world's population, the United States is by far the world's largest consumer of energy and the largest producer in terms of GDP. In reflecting on the points made in chapter 2 regarding energy inequality and security, this attribute is particularly troubling in the context of U.S. relations with the rest of the world. Because of its incredible energy needs, the United States relies heavily on foreign sources, and as a result, security issues are an important part of its energy policy. But U.S. energy issues are even more multifaceted. Other important issues include public land use, environmental health, and management of energy provision.

This chapter examines energy issues in the United States. First, there is a basic overview of energy facts and statistics. Then there is a brief history of energy and environmental policy. Finally, there is a discussion of current, relevant, U.S. issues.

Energy Facts and Statistics

This section examines basic facts and statistics of energy use in the United States. It provides an overview of energy resources that are utilized in the United States, where they are used, how much is consumed, and where they come from. The goal of this section is to present the reader with a background of how energy is produced, distributed, and consumed in the United States. The statistics described in this section were obtained from the U.S. Energy Information Administration (EIA). More information and the historical trends of these figures are graphically represented in chapter 6.

The United States is an incredibly energy-intensive nation. In 2004, it consumed an estimated 100.4 quadrillion Btus of energy, double the amount that was consumed in 1963 (EIA 2005a, v). This works out to be more than 340 million Btus of energy per person. Of this energy, approximately 21.2 percent was consumed in the residential sector, 27.8 percent in the transportation sector, 17.5 percent in the commercial sector, and 33.2 in the industrial sector.

Natural resources are the raw ingredients of the energy system in the United States. Fossil fuels provided the majority of the energy consumed in 2004 (86.3 quadrillion Btus), nuclear energy provided 8.2 quadrillion Btus, and renewable energy provided the rest (EIA 2006a). Much of the oil consumed came from other countries. In 2004, the United States imported 4.81 billion barrels of petroleum products (EIA 2006c). Approximately 2.01 billion barrels came from Organization of Petroleum Exporting Countries (OPEC) nations and the remaining 2.69 billion barrels from eighty non-OPEC nations (EIA 2006c). In the same year, the United States exported 383.6 million barrels (EIA 2006b).

Despite importing a large share of its oil, the United States had oil reserves estimated to be between 22 and 29 billion barrels in 2004 (EIA 2005c). Most of these reserves (17 billion barrels of oil) are located in twenty-four states and three offshore sites (one in the Pacific Ocean off the coast of California and two in the Gulf of Mexico located off the coasts of Texas and Louisiana).

Natural gas is increasingly being used as a utility in the residential sector for heating and cooking purposes. It is also used extensively in the industrial sector. Because it is less polluting, it has been promoted as a potential alternative to petroleum for transportation. Natural gas reserves in the United States are estimated to be around 204.4 trillion cubic feet (EIA 2005c). In 2006, the United States consumed 21.9 trillion cubic feet of natural gas and produced 18.8 trillion cubic feet (EIA 2005a, 186). The United States imported 4.3 trillion cubic feet and exported 854 billion cubic feet.

Most of the natural gas in the United States is produced from onshore wells. Texas, Louisiana, and Oklahoma are the three primary producing states, but wells are located in many other states (mostly in the western United States). The use of offshore wells is also increasing, the bulk of which are located in the Gulf of Mexico.

Coal is the main energy source used in electricity production in the United States. The country's estimated recoverable reserves of coal are 18.1 billion short tons (EIA 2005b). Because coal resources are abundant, most of the coal consumed is produced domestically. In 2004, of the 1,111.5 million short tons of coal produced, only 48 million short tons were exported. The United States imported 27.3 million short tons and consumed a total of 1,104.3 million short tons (EIA 2005a, 206).

Coal is mined in many states east of the Mississippi, where most of the coal reserves are anthracite. It is mainly extracted using shaft mining techniques. Since the 1980s, surface mines have been increasingly popular, and the production of coal west of the Mississippi (where most of the coal is located close to the surface) has increased dramatically. There are many reasons for this trend. One is that surface mining techniques are safer. Another is that the sulfur content in western coals is found to be much lower than in coals from the east. Lower sulfur content is an attractive attribute to electricity producers because it reduces the amount of SO_2 that is produced during coal combustion.

The utilization of nuclear energy has not lived up to the initial expectations of many energy analysts. Public outcry to the Three Mile Island scare caused a decrease in support for nuclear energy; because of the scare no new nuclear power plants have been built in the United States since 1979. In 2004, there were a total of 104 nuclear generators operating in the United States, generating a total of 788.6 billion kilowatt-hours of electricity (EIA 2005a, 276). The United States relies mostly on uranium imports to supply fuel for its nuclear generators. It purchased 66 million pounds of uranium oxide in 2004 (277).

Renewable energy sources comprise only a small part of the energy equation. These sources provide only 6 percent of the total energy consumed in the United States. Most of the renewable energy consumed is derived from hydropower (45 percent). Wood comes in second at 33 percent, and then waste (9 percent), geothermal (6 percent), alcohol fuels (5 percent), wind (2 percent), and solar (1 percent) (EIA 2005a, 280). Most of this energy is converted to electric power production. The industrial sector is the main user of renewable energy, followed by the residential, transportation, and commercial sectors (282). The transportation sector is one area where renewable energy use is dramatically growing as alcohol-based fuels are being developed.

Electricity is an important part of the energy system. The United States generated approximately 3.9 trillion kilowatt-hours of electricity in 2004 (EIA 2005a, 226). Electric utility companies produce most of the electricity in the United States, with the industrial and commercial sectors also generating some. Coal is the main source of energy for electricity generation (50 percent), nuclear provides 20 percent, natural gas generates 18 percent, and hydropower contributes 7 percent. The residential sector uses the greatest amount of electricity, purchasing 1.3 trillion kilowatt-hours in 2004 (267). The commercial and industrial sectors were not far behind, consuming 1.2 and 1.0 trillion kilowatt-hours, respectively. And recall that not all of the chemical energy contained in coal and nuclear fuels can be harnessed for energy services. A substantial amount of energy is lost as heat energy in generating electricity.

What is the point of all this information? It is easy to look at these numbers and become overwhelmed with the amount of data associated with energy use. However, these statistics are essential for describing energy dynamics. From them, one can understand how much and what kind of energy is used, where energy savings can be found, and what future energy trends might be. This information is also useful in making decisions about energy provision and distribution in society. However, as the next section describes, despite having loads of information, it has been difficult to formulate a comprehensive energy policy for the United States.

Energy and Environmental Policy

Energy policy in the United States is multifaceted. It integrates many issues that are important for both U.S. international relations and domestic environmental and natural resources policy. This section examines the history of U.S. energy and environmental policies. First under consideration is the progression of U.S. national energy policy over the past seventy years. Next is a description of nuclear energy policy. Finally, there is a discussion of environmental policies that have been developed to deal with energy issues.

National Energy Policy

Because so many different interests are involved in energy issues, it has been difficult for the United States to implement a comprehensive energy policy. Prior to the late 1940s, federal involvement in energy policy was limited to regulatory oversight of utility distribution (discussed later in this chapter) and intervention in energy supply during times of war. In World War I, a mandatory energy conservation program was necessary to overcome a fuel crisis that occurred from lack of foresight in coal production. During World War II, a rationing program issued gasoline coupons to U.S. motorists (Melosi 1985, 187). Subsidies for electricity infrastructure are another way that the federal government intervened in energy production. For instance, many of the large dams in the western United States were built using federal funds. The federal government also played a large role in developing nuclear energy policy, both through subsidies and regulations regarding the purchase of nuclear fuels (nuclear energy policy is discussed later in this chapter).

Despite these early interventions, the U.S. government generally let the market regulate energy provision unless an energy crisis occurred. After World War II, the federal government played a larger role in energy pricing, requiring petroleum import quotas for the purpose of protecting U.S. companies and promoting domestic energy resources. The Mandatory Oil Import Program (MOIP) was established in 1959 by the Eisenhower administration. It created a quota system for the amount of oil imports allowed into the country. Individual oil companies were issued licenses for their imports. The effect of the MOIP allowed domestic companies to keep their prices high and also reduced global demand for oil, resulting in a downward pressure on global prices (Kapstein 1990, 132).

Other than the federal influence on pricing and subsidies provided to particular industries, a long-term plan that addressed the energy needs of the entire country was elusive. The Truman, Eisenhower, Kennedy, and Johnson administrations failed to develop comprehensive energy plans. Instead, the pattern was crisis management substituting for a well-thought-out long-term program (Smith 2004, 150).

In 1971, Nixon eliminated the MOIP but maintained price controls on the petroleum market. When OPEC raised energy prices in 1973 and the subsequent oil embargo ensued, many analysts became concerned over the uncertainty in the energy markets. Nixon responded by initiating Project Independence, a plan to make the country energy self-sufficient by 1980 (Miller and Miller 1993, 21). The Emergency Petroleum Allocation Act (EPAA) was also passed. It established pricing controls for the purpose of equitable distribution of petroleum resources.

President Gerald Ford had difficulty in addressing energy issues. His administration pursued Project Independence and emphasized the role of national planning and federal control in energy issues. Although his plan embodied many of the same measures undertaken by the Nixon administration, he was unable to develop a unified approach to energy.

The Carter administration attempted to formulate a long-range, comprehensive energy program for the country. President Jimmy Carter's agenda not only addressed long-term concerns of energy availability, it sought ways in which the United States could maintain a secure energy future. In 1977, the Department of Energy Reorganization Act created the Department of Energy (DOE). The newly created agency consolidated the many energy programs and agencies of the federal government into one bureaucratic structure (Fehner and Holl 1994, 22). President Carter emphasized that energy goals needed to reduce energy demand, reduce dependence on foreign oil, and increase energy efficiency. The creation of the DOE was essential for the accomplishment of these goals. Carter also announced his National Energy Plan (NEP) in 1977, emphasizing the serious nature of the energy crisis by calling it "the moral equivalent of war." The president recognized that by ignoring the increasing scarcity of petroleum, the United States "would subject our people to an impending catastrophe." The NEP focused on energy conservation. It called for major improvements in energy efficiency for existing buildings and acceleration of the applications of solar technology. It also contained various conservation incentives, such as insulation credits, weatherization grants, energy audits, and loans for solar energy systems. It taxed gas-guzzling cars and prohibited the use of oil or gas in new electricity generation and new industrial plants and established voluntary electrical rate designs. Additionally, the plan called for a reduction in average annual energy growth to less than 2 percent; reduction in natural gas consumption by 10 percent; and continued reductions in imported oil (Smith 2004, 151). The Natural Gas Act was also an important part of the NEP. It is discussed later in this chapter.

One of the most important parts of the NEP bill was the Crude Oil and Equalization Tax (COET). It proposed raising oil prices over the next three years to encourage the application of energy efficiency measures and thereby reduce demand. The revenues from the tax were to be diverted into several government programs. The idea behind the COET was that price incentives were needed to encourage efficiency. Higher prices not only promote efficiency, they are also necessary to reduce energy demand and promote the development of better technologies (Smith 2004, 151; Nivola 1986).

Many parts of the National Energy Plan were very controversial. Northeastern congressional representatives with constituents dependent on home heating oil, as well as westerner representatives whose constituents used their automobiles for extended travel, denounced various parts of the plan that would have increased the price of energy (Smith 2004, 151). Ultimately, the NEP failed to present a pricing strategy. While it did succeed in promoting conservation strategies among the U.S. public, the most ambitious aspects, like the COET, failed to pass Congress.

A major piece of legislation that did pass during the Carter administration was the Public Utilities Regulatory Policies Act of 1978 (PURPA). PURPA required that utilities purchase electricity from independent generators. These generators, known as "qualifying facilities," utilized their waste heat to produce electricity, a process called cogeneration. By supporting facilities that recycled their waste heat, PURPA sought to encourage greater energy savings. In order to encourage cogeneration, PURPA allowed qualifying facilities to be exempted from state and federal regulations. At the same time, utilities that wished to purchase power from qualifying facilities were required to pay what the utilities would otherwise spend to generate or procure power. Hence, an incentive existed for power plants to install cogeneration technology.

Energy policy during the Reagan administration shifted away from conservation goals. Rather it focused more on the ability of free markets to satisfy current and future energy needs. One dramatic example of this occurred just after Reagan's inauguration when he removed all price controls on crude oil (Davis 2001, 153). Reagan never adopted a unified energy plan. Instead, the administration worked to limit enforcement of existing environmental laws and promote drilling and mining on federal lands. The idea was that if energy companies were granted access to the lands, and if prices were determined in a competitive market, energy concerns would be alleviated. Not only would the market provide cheap energy, it would promote innovation in energy efficiency measures. Federal government spending on research and development of alternative energy sources was cut dramatically. Although market forces did prove to be pragmatic and energy usage as a percentage of GNP declined from the adoption of energy efficiency measures, the reduction in renewable energy research caused greater reliance on fossil energies (Smith 2004, 152).

Following the Regan administration, attitudes toward federal control of energy provision became more moderate. The Comprehensive National Energy Policy Act (CNEPA) was signed by President George H. W. Bush on October 24, 1992. CNEPA promoted increased energy efficiency, established targets for decreasing oil consumption, implemented energy efficiency standards for federal buildings, and instigated measures to address global warming.

The legislation addressed efficiency concerns in the industrial, commercial, and residential energy sectors. Efficiency standards for lights, showers, toilets, faucets, small motors, and commercial heaters and air conditioners were to be met by federal buildings and public housing. Federal technical assistance and incentives, such as utility grants, were given to states to update their building codes. Grants were also available for industry to promote efficiency. Voluntary guidelines were issued for homes along with a mortgage pilot program for energy-efficient homes and retrofitting. CNEPA also created a director of climate protection to oversee greenhouse gas research and policy-making, foreign aid, and exploration of technology to combat global warming. It also targeted a 30 percent increase in energy efficiency by 2010, a 75 percent increase in the use of renewable energy sources by 2005, and a decrease in oil consumption from 40 percent of total energy use to 35 percent by 2005 (Idelson 1992).

Although improvements in energy efficiency standards were made, CNEPA was not entirely environmentally friendly. The legislation did not promote an energy tax increase (which would have decreased demand), nor did it increase gas mileage standards. In fact, the bill provided $1 billion in tax breaks to independent oil and gas drillers. Finally, despite the devastating *Exxon Valdez* oil spill in Prince William Sound, Bush promoted the opening of Alaska's Arctic National Wildlife Refuge (ANWR) for oil exploration. The ANWR provision was dropped from the final version of CNEPA, but it initiated a long battle over the rights to drill in the arctic wilderness area (discussed later in this chapter).

The election of Bill Clinton in 1992 increased hopes that energy policies would be connected to environmental goals. President Clinton publicly endorsed voluntary compliance in fuel efficiency standards and promoted greenhouse gas reductions. During his administration, the White House Office on Environmental Policy (OEP) was created with an agenda to stress the economics of environmentalism, including creating jobs and business opportunities for new technologies (Sullivan 1993). Additionally, research funding for renewable energy sources was dramatically increased. Despite these measures, the Clinton administration failed to link energy policy with other issues such as conservation and environmental protection. Furthermore, Clinton was initially unable to formulate a comprehensive plan for energy policy. While an additional gasoline tax of 4.3 cents a gallon was enacted into law, it was part of an economic stimulus package and was not expected to affect consumption to any significant degree (Smith 2004, 152).

In the spring of 1998, the Clinton administration released the Comprehensive National Energy Strategy intended to provide long-term guidance for the country's energy needs. The five goals of the strategy sought to improve energy efficiency; ensure against energy disruptions; promote energy production and use in ways that respect health and the environment; expand energy choices; and enhance international cooperation (DOE 1998, viii). Specific initiatives were outlined that promoted the development of alternative fuel vehicles, promotion of natural gas over coal and oil, and a proposed mandate requiring that electric utilities obtain at least 5.5 percent of their power from renewable resources by 2010.

Electricity regulation was also an important issue in the 1990s. During this time, state governments sought to increase competition among power companies by either repealing or otherwise disabling the restrictions PURPA placed on electric utilities and how they obtain their power (Smith 2004, 153). Critics of PURPA requirements contend that the regulations interfere with competition in the energy market, resulting in higher prices for the consumer. The question of energy regulation is an important issue in U.S. energy policy, and it is discussed in greater detail later in the chapter.

The George W. Bush administration moved away from the goals of Clinton's energy strategy. In many ways, sustainable energy measures were abandoned in favor of further promoting fossil energies. Vice President Dick Cheney promoted the position of the Bush administration well when he announced in 2001 that the main energy problem that the nation will face is one of supply (Kahn, 2001a). The vice president did not mention reducing demand, only increasing supply. The focus is on technological fixes, with the more immediate concern of increasing traditional energy supplies, including obtaining more energy from the U.S. Arctic. Energy plans drafted by the Bush administration also promote tax breaks to companies that seek to increase domestic production of oil and natural gas.

The attention and incentives given to increasing supplies of fossil fuels has prompted criticism that the Bush administration is dominated by industry interests. This suspicion was sparked early in the administration with the formation of the Energy Task Force headed by Cheney. The main goal of the task force was to find ways to secure the energy supply for the United States. When the task force released its report, the General Accounting Office (GAO) requested that Cheney release information regarding the executives that advised the energy panel in closed-door meetings (Kahn 2001b). Cheney refused, and the GAO sought a court order to obtain the documents. A federal judge ordered the release of documents in 2002. It was uncovered from those documents that the Energy Task Force had met with eighteen of the energy industry's twenty-five top contributors to the Republican Party (Van Natta and Banerjee, 2002).

In August 2005, President Bush signed the Energy Policy Act of 2005. This legislation promoted a mixed approach to energy policy that on one hand expanded the use of biomass energies and biofuels and provided incentives for efficiency gains, but on the other hand provided subsides for oil and gas development. The bill also made dramatic changes to electricity regulation (discussed later). Furthermore, it focused on increasing the use of nuclear energies by granting an estimated $7 billion in subsidies via an electricity production tax credit (UCS 2005). The Bush administration has shown strong interest in enhancing nuclear capabilities, but as the next section shows, it has been difficult to promote nuclear energy in U.S. society.

Nuclear Energy Policy

The development of nuclear energy policy deserves special mention because it has been guided by the federal government since the initial discovery of nuclear power. The government has subsidized much of the research and development of nuclear technology, has controlled the allocation and distribution of nuclear fuel, and has established various programs for the storage of radioactive waste. In short, nuclear energy is unique because the entire fuel cycle is regulated by the federal government. This section describes the various nuclear energy policies that have been enacted.

Nuclear power was first developed for military purposes. The detonation of atomic bombs in the cities of Hiroshima and Nagasaki in 1945 demonstrated the power of nuclear weapons to the world. The two bombs killed over 340,000 people either from the direct blast of the bomb or from radiation exposure (Fehner and Holl 1994, 11). The devastation not only changed the nature of international relations, it also demonstrated the necessity for government control of nuclear power. In order to promote peaceful uses of nuclear energy and to ensure that international control over nuclear energy was maintained, the United Nations Atomic Energy Commission (UNAEC) was created in 1946.

With domestic nuclear policy, the United States sought to maintain control over its atomic energy monopoly. In 1946, President Truman signed the Atomic Energy Act. This bill created two government agencies, the Atomic Energy Commission (AEC) and the Joint Committee on Atomic Energy (JCAE). The AEC was given ownership of all nuclear materials and reactors in the United States. It acted as a regulatory agency for radioactive materials, granting licenses to facilities generating, using, and researching nuclear energy. The JCAE was established as a congressional information and watchdog agency to oversee the nuclear activities of the AEC.

In 1954, the United States sought to enhance development of the nuclear power industry. The 1954 Atomic Energy Act promoted private nuclear development by granting subsidies for research and development of nuclear reactors. It stipulated that private companies could own nuclear reactors while the federal government retained ownership over nuclear fuels. The goal was to motivate a greater private interest in nuclear energy and to ensure the public that nuclear energy would provide the United States with cheap, abundant sources of electricity for many generations.

Despite these efforts, support for nuclear plant construction remained low. Nuclear energy was still very expensive to develop, and the prevailing idea in industry was that nuclear power would cost more to produce than could be gained from profit. Additionally, the potential damage that would result from a nuclear accident substantially increased the liability for private operators of nuclear power plants. In an effort to address the liability issue, Congress in 1957 passed the Price-Anderson Act limiting the liability of individual companies. The Price-Anderson Amendments in 1988 raised the liability limits from $5 million per facility per incident to $63 million per facility per incident. Furthermore, plant operators (or licensees) were not required to pay out more than $10 million in any one year in case of liability under the act (Smith 2004, 149). Although liability issues were addressed by the Price-Anderson Act, nuclear power only became a viable option when electricity shortages and environmental concerns shifted interest away from the coal-fired power plants. Consequently, the 1960s saw a dramatic growth in the nuclear power industry.

The 1970s were a difficult decade for the development of nuclear power. The Ford administration significantly changed the structure of federal energy agencies with the Energy Reorganization Act of 1974. This legislation dissolved the Atomic Energy Commission and created the Energy Research and Development Administration (ERDA) in its place. The ERDA took over the research and development of all energy forms, and the Nuclear Regulatory Commission (NRC) assumed the AEC's regulatory function over the nuclear industry. The 1974 act proved to be difficult for the nuclear power industry, and demand for reactors dropped after 1975 (Melosi 1985, 308). The organizational changes of the 1977 Energy Reorganization Act also undermined nuclear promotional campaigns and placed the NRC under the administration of the newly established Department of Energy (DOE). A final devastating blow to the nuclear power industry occurred with the Three Mile Island (TMI) accident in 1979. After the accident, all existing orders for reactors were canceled, and as of 2006, none have been placed since in the United States.

In the 1980s, safety and waste issues dominated the nuclear energy agenda. The Nuclear Safety Research, Development, and Demonstration Act was passed in 1980 largely as a response to the TMI accident. This legislation sought to improve the safety of existing nuclear power plants. It mandated that standards be established for the construction of nuclear facilities and ensured that safety rules be implemented in the United States' nuclear power plants. Conflict over nuclear waste storage also became a heated issue in the 1980s as regulators sought agreement on site feasibility for long-term storage. The Nuclear Waste Policy Act in 1982 mandated that geologic disposal was the solution for nuclear waste storage, and federal agencies were charged with the task of finding suitable locations for nuclear waste repositories, one in the East and one in the West (Long and Ewing 2004). In 1987, this ruling was revised to only one central storage location, as it proved impossible to locate an appropriate site in the East. Nuclear-storage issues are discussed later in the chapter.

Despite the setback of the 1970s, the pursuit of nuclear power was not eliminated. Concerns over increasing reliance on foreign energy sources and global climate change have caused President George W. Bush to rekindle efforts to develop nuclear resources. The Bush administration promoted nuclear energy as a means of addressing the problem of climate change. Abandoning the Kyoto Protocol was one particularly glaring consequence of the administration's approach to energy use. Stating it would be devastating to the economy, the administration asserted that it was not fair that developing nations like India and China would not have to meet the same requirements as the United States. The nuclear option was offered as an alternative approach for reducing greenhouse gas emissions. It remains to be seen whether the energy issues associated with nuclear power will be resolved.

Environmental Regulation

Chapter 2 discussed the problems of air and water pollution and land degradation that result from fossil fuel combustion. Throughout most of the twentieth century, there have been legislative and regulatory efforts in the United States to minimize the impact of energy industries on the environment. This section provides a brief overview of the laws that govern how the United States deals with environmental problems.

Environmental concerns associated with energy use were first addressed in the early 1900s. In urban areas, smoke abatement coalitions were formed to raise public awareness about air pollution problems associated with industrial activities. Other environmental efforts sought to reduce pollution from oil industries. The process of drilling and refining produced sludge waste that would pollute aquatic ecosystems. Additionally, the practice of gas flaring, used to remove unwanted gas from crude oil, endangered both human health and the environment. The 1924 Oil Pollution Control Act was developed to control pollution that was emitted during the extraction and production of petroleum products. Although it developed standards for handling sludge waste, oil vessel effluent, and gas flares in the process of oil drilling and refining, this legislation did not provide the necessary enforcement to ensure pollution abatement.

Pollution issues were not seriously addressed until the latter half of the twentieth century. Throughout the 1950s and 1960s, negative environmental consequences associated with energy consumption were apparent as air and water pollution became more visible. The 1969 oil spill off the coast of California raised public awareness about the negative impacts of oil spills. In 1975, a total of 12,000 oil spills were reported in the United States (Melosi 1986, 299).

The United States became a world leader in the area of environmental regulation when the National Environmental Policy Act (NEPA) was passed in 1969 and the Environmental Protection Agency (EPA) was formed in 1970. NEPA required that an Environmental Impact Statement (EIS) be prepared for potentially environmentally damaging activities. Other environmental laws specified standards for air and water pollution, land degradation, and wildlife protection. Overall, the 1970s were a pivotal time period for environmental legislation. Although they have been revised many times, these laws serve as the foundation for how the United States addresses the negative impacts from energy use.

Since a large amount of environmental damage occurs from the use of fossil fuels, the United States has developed extensive air pollution laws to limit the various pollutants emitted during fossil fuel combustion. The following discussion of air pollution illustrates how environmental regulation can change over time.

For instance, the Clean Air Act was originally passed in 1963 and amended in 1967, 1970, 1977, and 1990. Not only do these laws reflect prevailing interests in the regulatory bodies of government, they also became more nuanced as more information became available and new technologies were adopted.

The issue of air pollution was initially considered to be a state regulatory responsibility. Federal intervention began in 1955 when Congress offered technical expertise and financial assistance to the states to reduce air pollution. The Clean Air Act of 1963 empowered federal officials to intervene in interstate air pollution matters only at the request of state governments. Under this law, the apparatus for enforcing pollution abatement was so cumbersome that between 1965 and 1970, only 11 abatement actions were initiated under the 1963 Clean Air Act (Rosenbaum 1973, 153). The Air Quality Act of 1967 further addressed the problem by mandating the secretary of the Department of Health, Education, and Welfare to establish Air Quality Control Regions (AQCRs), where states within each region were mandated to establish air quality standards. Because they deferred enforcement to the states, these early acts were largely ineffective at mitigating air pollution. Not only did state governments lack the resources needed to enforce these laws, there was no incentive for them to do so. Additionally, air pollution regulations were not consistent from state to state. National air quality standards were necessary in order for air quality to improve uniformly across the country.

The Clean Air Act of 1970 and subsequent amendments in 1977 and 1990 represented significant increases in federal involvement in air pollution regulation. The act directed the Environmental Protection Agency administrator to establish "national ambient air quality standards" (NAAQS). The NAAQS defined set limits of pollution that could be emitted for six primary pollutants: SO_2, NO_X, ozone, carbon monoxide, particulate matter, and lead. It also required uniform national standards for "hazardous" air pollutants, or those known to severely impact human health. Under the 1970 act, states were required to develop state implementation plans (SIPs) that described their implementation of NAAQS. The act was to be federally enforced by the newly established EPA.

The Environmental Protection Agency created a number of classification schemes in its implementation of the provisions of the Clean Air Act. Emission sources are divided into "major" or "minor" and are found either in "attainment areas" (where air quality goals are met) or in "nonattainment areas" (where air quality does not meet standards). Standards were developed for both stationary (e.g., power plants) and mobile sources (e.g., vehicles) of air pollution. Sources can be "new" or "existing," depending on if the source of emission was in place when permits were first issued. Regulations vary, taking into account whether a source is major or minor, in an attainment area, is stationary or mobile, or new or existing (Smith 2004, 91).

The 1977 amendments to the Clean Air Act placed Air Quality Control Regions into specific classes that detailed how much air pollution could be emitted in a specific area. They also established emission standards for automobiles and trucks requiring a reduction in carbon monoxide and hydrocarbon emissions. The amendments also provided that federal implementation plans (FIPs) could be imposed on a region if SIPs had not been submitted to the EPA. As a result of several lawsuits by environmental organizations, the EPA was forced to develop air pollution FIPs in Phoenix, Arizona; the metropolitan Los Angeles area; Bakersfield, Sacramento, Ventura, and Fresno, California; and Chicago, Illinois.

The 1990 Clean Air Act Amendments (CAAA) further revised regulations for six criteria pollutants: ground-level ozone, carbon monoxide, sulfur dioxide, nitrogen dioxide, lead, and particulate matter. It clarified "nonattainment" status and developed classification schemes for three pollutants, ozone, particulate matter, and carbon monoxide, based on their level of severity. These classifications specified increasingly stringent abatement measures. Stationary sources were redefined according to the nonattainment level of their location. For areas that had not attained ambient air quality levels, SIPs were required to specify abatement measures. The 1990 CAAA also established an operating permit program, which required sources to provide information on which pollutants are being released, in what quantities, and to provide plans to monitor levels of pollution. Other parts of the 1990 CAAA deal with air toxics, protecting stratospheric ozone, and acid rain.

One innovation that is particularly noteworthy of the 1990 CAAA is the use of cap-and-trade programs to reduce SO_2 and NO_x emissions. Cap-and-trade programs are market-based mechanisms that cap emissions at a certain level and issue permits to companies that specify how much of a certain pollutant they are allowed to emit. If a company exceeds its allowable emissions, it has the opportunity to buy pollution credits from another company that emits less than its allowable limit. Hence, a cap-and-trade system turns pollution into a commodity by allowing companies to trade their pollution permits. The 2005 Clear Skies legislation expands the use of cap-and-trade programs. The bill still has yet to be enacted, but if the legislation passed, it would reduce the caps of allowable emissions for NO_x and SO_2. The 2004 Mercury Rule implemented a cap-and-trade system for mercury emissions (USEPA 2005). The mercury provision is controversial because mercury is listed as a hazardous air pollutant (HAP). Critics note that previously it was illegal to implement a trading system for emissions that are categorized as HAPs.

Air pollution was not the only concern addressed in the 1970s. Environmental issues associated with water bodies were addressed with legislation. The Clean Water Act (CWA) was passed in 1972. It created a system for monitoring point source pollution (i.e., pollution that is emitted from a known source and can be measured). It established standards for certain effluents and required that the best available pollution control technology be used to remove pollutants from water that is discharged into the environment. It established a National Pollution Discharge Elimination System (NPDES) permit program that requires polluters to obtain a permit for discharge. Although the CWA applied to all polluting industries, it affected the energy companies because it established a system for regulating much of the waste that was produced in water effluents during the process of energy production.

Important regulations for land management were also established during the 1970s. The 1976 Federal Land Policy and Management Act (FLPMA) mandated that the Department of the Interior (DOI) outline land use plans for over 488 million acres of land (Davis 2001, 147). It specified coordination among different land management agencies and required careful consideration of environmental issues in long- and short-term planning initiatives. Although the FLPMA was not specifically designed to address energy issues, it had implications for the various extractive industries that sought to develop energy on public lands.

The 1977 Surface Mining Control and Reclamation Act (SMCRA) was a significant land management law for the coal industry. It mandated that coal industries reclaim land after mining operations ceased, and it established a revolving fund for the purpose of reclaiming abandoned mine sites. The legislation also created the Office of Surface Mining (OSM) to manage the regulatory duties that SMCRA required.

The coal industry was also affected by the 1969 Federal Mine Safety and Health Act, which sought to improve labor conditions in mines by implementing mandatory safety standards to be enforced by mine operators and followed by miners. The goal of the law was to reduce the number of fatal mining accidents and risks of exposure to toxic chemicals. It was revised in 1977. The Mine Safety and Health Administration (MSHA) was established to enforce safety standards and educate the mining industry, both miners and operators, about the new regulations.

In addition to land management and air and water pollution laws, there are several other federal laws that impact the energy industry. Waste disposal laws, such as the Resource Conservation and Recovery Act (RCRA) (mandates how waste should be disposed) and the Comprehensive Environmental Response, Compensation and Liability Act (CERCLA) (dictates rules for the cleanup of hazardous waste sites), have had important implications for energy production. These regulatory laws specify how toxic and hazardous waste produced by the drilling, mining, and refining industries should be disposed of and how contaminated sites should be cleaned.

The environmental legislation passed during the 1970s in the United States was monumental in its acknowledgement of environmental problems associated with energy use. Although environmental regulations are not specifically a part of energy policy, they must be considered by facilities that extract, produce, and distribute energy resources. Regulations need to be revisited and revised regularly in order to adapt to the changing trends in energy industries. Even when environmental and energy legislation is implemented, issues arise with how those laws are interpreted and the appropriate measures that need to be taken to address new problems. The next section addresses issues that are important for U.S. society.

U.S. Energy Issues

The previous sections on U.S. energy and environmental policy demonstrate how incredibly difficult it is to develop and implement a comprehensive national energy policy that incorporates these components. Energy issues are dynamic and evolving. There are not always clear-cut strategies for managing the various components involved in energy production and distribution.

When new energy issues arise, it is often difficult to implement policies that satisfy the various interests involved. This section addresses energy issues important to U.S. society. First it examines debates over energy development on federal lands. Then it examines the problem of electricity regulation.

Energy and Federal Lands

One of the most important energy issues in the United States is the use of federal lands for purposes of energy development and nuclear waste storage. Federal lands consist of more than 700 million acres of land owned and managed by federal government agencies. There are four major federal land management agencies: the Bureau of Land Management (BLM) (465 million acres of land used mostly for mining and ranching); the U.S. Forest Service (USFS) (187 million acres of forestlands managed for multiple uses); the U.S. Fish and Wildlife Service (USFWS) (26.5 million acres of varied lands where wildlife populations are managed); and the National Park Service (NPS) (23 million acres comprised of national parks and monuments preserved for their unique characteristics and beauty) (Clarke and Angersbach 2001, 35). The federal government has also designated national wildlife refuges, wild and scenic rivers, and wilderness areas for the purpose of preserving wildlife habitat and pristine areas from development. In addition, Indian reservations are considered to be federal lands, but often the tribes manage their own lands.

Despite who manages the land, these areas belong to the American people. They do not belong to any one entity, but are managed for the benefit of the public. Because of this, many issues are raised regarding how the land can be used and who can profit from its development. The Mineral Leasing Act (MLA) of 1920 gave energy companies the opportunity to lease land from the federal government for oil and gas drilling. The federal government collects royalties from the leases, which amount to a substantial revenue source for the U.S. Treasury (Laitos and Tomain 1992, 271). Between 1982 and 2002, the Minerals Management Service collected over $127 billion from oil and gas royalties (MMS 2003). Land managed by the BLM and the Forest Service can be opened for leasing, but federal lands that are designated as national parks and monuments, national wildlife refuges, wild and scenic rivers, and wilderness areas cannot be leased for drilling. Offshore drilling activities are also leased from the federal government pursuant to the 1953 Outer Continental Shelf Lands Act (OCSLA), a law that gave the Department of the Interior management responsibility over offshore drilling leases. Coal resources can also be mined on federal lands, but they are leased according to provisions defined in the 1976 Federal Coal Leasing Amendments Act.

Drilling in the Alaskan National Wildlife Refuge and the designation of Yucca Mountain in Nevada as a nuclear waste repository site are two examples of issues that concern the use of federal lands for energy development and nuclear-waste storage.

Arctic National Wildlife Refuge

As fossil fuel resources become increasingly scarce, and because energy security in the United States is threatened by its reliance on foreign energy sources, the federal government has expressed greater interest in promoting energy development on federal lands. As noted above, much of this development occurs on land leased from the federal government by resource extractive industries. However, the debate is increasingly being centered on whether or not to allow energy development on federal lands that were set aside for conservation purposes: lands that are located in national parks, monuments, wilderness areas, and wildlife refuges. While energy development on federal lands is being debated for areas all over the country, development of the Arctic National Wildlife Refuge (ANWR) is the most publicized and contentious of these discussions.

The issue of oil development along the North Slope in Alaska began in 1968 when oil was discovered in Prudhoe Bay. Soon after the discovery, oil companies announced plans to develop the Trans-Alaska Pipeline System (TAPS), a forty-eight inches in diameter pipeline that would transport crude oil over 800 miles from Prudhoe Bay to Valdez (Gorman 2001, 290). The development of TAPS, however, was a difficult matter. The pipeline would need to transverse federally owned land over fragile stretches of permafrost. Issues of habitat fragmentation for caribou migrations and questions of what would happen to the landscape in the event of pipeline damage raised concerns that required resolution for the proposed pipeline. Additionally, public attitudes toward pipeline development were wary as images of the Santa Barbara oil spill were burned into Americans' memories. Many preservationist and environmentalist organizations in the United States rallied against the pipeline and sued the government for failing to follow newly enacted NEPA guidelines in its overview of the proposed pipeline. Additionally, they contended that the area of land to be used for pipeline operation exceeded the allowable limit of land under the 1920 Mineral Leasing Act (291).

In response to these criticisms, the DOI completed two drafts of an Environmental Impact Statement that addressed the environmental concerns raised. By 1972, the U.S. District Court of Appeals ruled the federal government had met NEPA guidelines, but stated that Congress needed to revise MLA guidelines before pipeline construction would be allowed. Although the vote in Congress was close, energy concerns that arose over the 1973 oil embargo provided enough impetus to vote for the pipeline. Construction began in 1974. As a result of the EIS, several modifications were made to the original design, including 400 miles of elevated pipeline to protect permafrost and short segments of buried, refrigerated pipeline to accommodate caribou migrations (Gorman 2001, 294). Additionally, the pipeline was equipped with check valves to monitor for the occurrence of leaks. Oil first arrived in Valdez from the pipeline in 1977.

Energy development in Alaska grew after the construction of the pipeline, and by 2001 eighteen fields in the North Slope were producing oil (Streever 2002). Oil drilling and production are mostly conducted on Alaska state lands, but industry expansion has been proposed for surrounding lands to the east, toward the Arctic National Wildlife Refuge (ANWR), and for offshore operations into the Beaufort Sea.

ANWR is a 19-million acre stretch of land located along the eastern region of the North Slope. The refuge was established by the Eisenhower administration. Although it was designated as a wilderness area, wording in the designation did not preclude future oil and gas development. It is by far the largest wildlife refuge in the United States. It is the summer home of many species of migratory animals. An estimated 180 species of migratory birds and 129,000 porcupine caribou migrate to the refuge to breed in the summer (Cronon 2001). Bowhead whales also thrive off the coast of the North Slope. The animals are important for the livelihood and spirituality of the Gwich'in and the Inupiaq people who inhabit the area.

Since the 1990s, energy development in ANWR has been one of the most controversial energy issues in the United States. Development in the refuge was first considered in the early part of the decade, when oil concerns were heightened with the start of the Persian Gulf wars. Environmentalists fought the development of ANWR and succeeded in obtaining a ban on drilling in the area with the 1992 Energy Policy Act. The issue surfaced again during the George W. Bush administration, which made it a fundamental part of its energy plan.

The Bush administration has argued that as energy needs in the United States grow, the oil that lies underneath this wildlife refuge will be integral for maintaining supply. Furthermore, arctic drilling has been promoted for its importance to national security. Soon after September 11, 2001, the administration appealed to Congress to open ANWR to drilling, emphasizing the need to pass legislation that would allow the United States to rely less on sources of foreign oil. After meeting with his cabinet a month after the terrorist attacks, Bush told the press, "The less dependent we are on foreign sources of crude oil, the more secure we are at home" (Seelye 2001).

Environmental groups are strongly opposed to drilling in ANWR. Such groups often cite a study performed by the U.S. Geological Survey (USGS) that found oil development would "most likely" restrict the calving grounds of the caribou as well as result in higher calf mortality rates and weight reductions in both pregnant females and calves (Verhovek 2002). The Bush administration has dismissed the report as being based on outdated drilling practices. To alleviate environmental concerns, the administration asserted that technology has made exploring, drilling, and transporting oil more efficient and less damaging to the environment. Mark Pfeifle, a spokesperson for the Department of the Interior, noted that the report "demonstrates that with new technology, tough regulations and common-sense management, [the United States] can protect wildlife and produce energy" (Revkin 2001).

Concern for wildlife is not the only criticism that environmental groups have expressed. They also contend that the costs for drilling in the refuge far outweigh the benefits. The amount of oil that could be recovered from the refuge is estimated by the USGS to be between 5.7 and 16.0 billion barrels, with a mean average of 10.4 billion barrels (USGS 1998). This small amount of oil is only enough to support the U.S. fuel supply for one year at best. Critics also note that ANWR is only a distraction from the other drilling sites that Bush has proposed. The energy policy recommends more than fifty new drilling areas in the western United States, most of them in the Rockies and some in national parks (Seelye 2002). It also grants tax breaks to oil and gas companies for greater production from offshore and marginal wells and provides money for coal mining technology.

As of 2006, the proposal to open ANWR for drilling had not succeeded in gaining congressional approval. Because of contentious opposition, the measure was not included in the 2005 Energy Policy Act.

Yucca Mountain

The location of a nuclear waste repository site has also been a contentious energy and land use issue. Currently the United States has over 40,000 tons of spent nuclear fuel and over 400,000 cubic meters of high-level radioactive waste located in 100 different sites in forty-two states (Long and Ewing 2004). Although this issue has plagued the federal government for five decades, the terrorist attacks of September 11 heightened concerns over the security of these facilities, the danger in transporting radioactive waste to various different sites, and the cost of storing nuclear waste in several different locations.

As noted in an earlier section, the Nuclear Waste Policy Act (NWPA) of 1982 mandated that the safe disposal of nuclear waste was the responsibility of the federal government. In 1987, an amendment to the NWPA established that waste disposal needed to be centralized in a single location, Yucca Mountain. Studies on the feasibility of Yucca Mountain as a potential site began in 1978. It is located in a remote region approximately 100 miles northwest of Las Vegas, Nevada, at the edge of the Nevada Test Site, an area that supported hundreds of nuclear tests. Yucca Mountain was chosen because feasibility studies demonstrate that the regional dry climate and its geologic stability are conducive conditions for waste storage. Most important, many studies show that the underlying water table is extremely deep and the mountain is situated in a closed water basin, meaning any water that flows into the area will not leave (OCRWM 2006).

One of the major problems with radioactive waste is that it persists in the environment for incredibly long periods of time. For example, the half-life of plutonium is estimated to be 24,000 years (Long and Ewing 2004). Throughout this time, the waste emits radiation that can be hazardous to human health. It also creates heat as it decays, making waste sequestration a difficult issue because storage conditions change over time. Due to the persistent nature of radioactive waste, public health and safety standards must consider extremely long time periods. In order to comply with the 1987 NWPA, the EPA developed a 10,000-year compliance period for determining public health and safety standards. Presidential authorization for the Yucca Mountain storage facility was finalized when President George W. Bush signed House Joint Resolution 87 in 2002.

Controversy in selecting one nuclear waste repository is inevitable. By doing so, the federal government made one state the nuclear garbage dump for the country. Although the Yucca Mountain site is on federal land, the repository imposes a burden on state government. Several issues must be addressed. How does Nevada deal with nuclear waste transport and what disaster plans are necessary should an accident occur? What types of implications does this selection have for growth in Las Vegas, Nevada's largest city? How does the state deal with negative publicity associated with nuclear waste? Although these concerns are often brushed off as being a not-in-my-backyard-syndrome (NIMBY), they are very legitimate logistical concerns that need to be considered.

Some critics say Yucca Mountain may not be as safe as the feasibility reports have determined. Scientists note that Yucca Mountain does not satisfy all of the technical conditions determined for safe geologic storage (Long and Ewing 2004). Furthermore, predicting geologic stability for extremely long periods of time is difficult. Changes to the Earth's hydrology in a particular area are not easy to predict, and Yucca Mountain could become an unsuitable repository over time. Others have criticized the federal government for determining Yucca Mountain to be the most suitable site before all the necessary studies were completed. Once the site was chosen, it was difficult to change course. Finally, critics declare reprocessing is a viable option for drastically reducing the amount of nuclear waste that needs to be stored. If the government were to fund reprocessing projects, perhaps a geologic repository would not be needed.

As of 2007, the Yucca Mountain Project has been stalled. In 2004, the D.C. Circuit Court of Appeals upheld the selection of Yucca Mountain as the nation's nuclear waste repository, but rejected the 10,000-year compliance period after hearing the case *Nuclear Energy Institute Inc. v. Environmental Protection Agency* (D.C. Cir. 2004). This ruling extended the project's timeline for development. In order to comply with the court's ruling, the EPA must develop new standards that take into account the findings of the National Academy of Sciences (NAS). The NAS concluded that the 10,000-year compliance period is not sufficient. Exposure periods are likely to occur after 10,000 years and a 1 million-year compliance period would be more consistent with available scientific evidence (Reblitz-Richardson 2005).

Utility and Electricity Regulation

Reliable electricity and utilities are important in every aspect of life, from the gas used to cook with to the electricity that lights homes and powers refrigerators. Because of this reliance, the distribution of electricity and natural gas is an extremely important component of the energy cycle in U.S. society. The regulation of electricity and utility distribution provides an interesting examination of federal and state interactions and how government regulation can impact the market for electricity and natural gas commodities. This section describes the development of utilities regulation in U.S. history. The electricity and natural gas industries sometimes are collectively referred to as "utilities."

Before the history and development of utilities regulation is examined, it is important to understand what a monopoly is and how monopolies create problems for market-based economies. Market-based systems operate on the basic foundations of supply and demand, which interact with one another to create the most efficient price structures. For example, if a supplier is charging too much money for a commodity, the demand for that good or service will decrease, forcing the supplier to lower the price in order to sell the good or service. Hence, the price of commodities changes according to shifts in supply and demand. A free market maximizes the efficiency of supply and demand because all participants have the equal opportunity and ability to buy and sell goods and services. In free markets, competition induces innovation, and the production of goods and services becomes increasingly efficient, provided there are plenty of players in the market.

A monopoly is described as a market failure. It occurs when there is only one or a few providers of a good or service that consumers can choose from. In society, natural monopolies exist because large amounts of capital funds are required in order to provide certain goods and services. For example, natural gas pipelines and electricity distribution systems cost a lot of money to construct and maintain. Additionally, they are most efficient when designed to service large numbers of people. Because of such large initial costs, efficiency is achieved when only one or two providers are operating on the market. This situation precludes everyone from being able to participate in the market and presents a problem for consumers. If only one or two providers are available for a necessary good and prices are dramatically increased, the consumer does not have the choice to switch to an alternate provider. Government regulations have been implemented as a solution to natural monopolies. By requiring price caps for certain industries, governments can ensure that prices for necessary goods and services are not dramatically increased.

In the early twentieth century, natural gas industries were regulated by state public utility commissions (PUCs). The 1938 Natural Gas Act established the Federal Power Commission (FPC) (which later became the Federal Energy Regulatory Commission [FERC]). The FPC was responsible for regulating interstate sales of natural gas, but not those of intrastate gas. Soon after the act became law, several issues became apparent. Since natural gas distribution operated on two different markets, interstate providers could not operate in intrastate markets and intrastate providers could not operate in interstate markets. The distribution restrictions created artificial shortages. For example, a natural gas shortage would occur in an intrastate market even though interstate providers had plenty of gas to sell; federal regulations prevented interstate providers from supplying gas services to consumers in intrastate markets. Pricing issues were also a concern, as consumers in intrastate markets were charged under a different structure than those serviced by interstate providers, whose prices were subject to regulation by the federal government. These issues created tensions between two groups of people. The first sought to deregulate the market, claiming that the natural gas industry was competitive enough to avoid the dangers of a monopoly. The second group was wary of deregulation, citing concerns over the effects that a free natural gas market could have on consumers.

Regulatory changes in the 1970s attempted to address the issues that arose from the 1938 law. Most significant, the Natural Gas Act of 1978 eliminated the distinction between inter- and intrastate markets and resulted in the federal government having a stronger role in the pricing of natural gas. The 1978 legislation also revised pricing structures for natural gas markets that ultimately sought to eliminate price controls (Laitos and Tomain 1992, 500).

The authorization of PURPA (discussed earlier) also affected the design of gas rate structures. Throughout the 1980s, FERC sought to adapt to the market with several regulatory changes. Despite the regulatory changes, many utility companies have been pressuring federal and state governments for deregulation.

Electricity regulation in the twentieth century experienced a similar trend of increasing federal involvement. In the 1920s, utility holding companies dominated the electricity market. These companies did not actually produce or distribute electricity; rather they consolidated and acquired control over smaller electricity providers. The smaller providers gained the capital funds required for electricity distribution, and the utility holding companies made a share of their profits. Unfortunately, this system was prone to abuse as stock manipulation and profiteering artificially bolstered the market (Melosi 1995, 119). Ultimately, the setup was detrimental to the electricity consumer, who experienced drastic and unpredictable price fluctuations.

In an effort to break the monopoly of utility holding companies, the Federal Power Act of 1935 was passed. The legislation partially alleviated the monopoly problem by allowing FERC to regulate interstate electricity rates while states provided their own rate structure for intrastate markets. The Public Utility Holding Company Act (PUHCA) of 1935 abolished the utility holding companies that comprised several levels of management (i.e., pyramid companies) and allowed the Securities and Exchange Commission (SEC) to investigate and regulate the business transactions of holding companies.

The 1978 enactment of PURPA was another important law affecting the electricity industry. It granted FERC the authority to intervene in the functioning of transmission lines. For example, it gave FERC the authority to order electricity companies to distribute electricity over another company's transmission lines (a process know as "wheeling") if such an action would promote conservation and efficiency (Laitos and Tomain 1992, 512). The law also increased FERC's jurisdiction in the area of utility rate-making. PURPA also required electricity generators to develop plans for electricity shortages. Finally, as discussed above, PURPA promoted energy conservation by mandating that electricity be purchased from cogeneration facilities.

The 1990s saw other changes proposed in the regulation of energy. State governments sought to deregulate their electricity markets in an effort to increase competition among power companies by either repealing or otherwise disabling PURPA regulations. Proponents of deregulation contended that by releasing utility companies from restrictions specifying what types of power they can utilize (e.g., cogeneration) and by allowing rates to be set by the market, prices would be more competitive and electricity consumers would have more choice. Opponents of the deregulation move contended that increased competition would foster a market situation oriented toward short-term economic returns rather than the long-term sustainability of renewable power resources.

The major impetus behind the deregulation debate has been the proliferation of large independent power producers who are not considered qualifying facilities under PURPA. It is argued that the tax advantages given to qualifying facilities unfairly discriminates against larger providers and stifles competition (Smith 2004). Additionally, consumers would likely benefit financially from the lower prices that could result from increased competition. The perceived advantage of the consumer being able to choose his/her own source of power, rather than being at the mercy of a single utility company, is one of the forces at the center of the deregulation debate.

The State of California experimented with utility deregulation. In 1996, Governor Wilson signed a law restructuring California's energy markets to allow for competition in electricity generation in order to drive costs down. Flaws in the deregulation plan ultimately led to a power crisis in California beginning in June 2000 when prices skyrocketed and supply plummeted, causing consistent rolling blackouts. Rolling blackouts occur when utilities do not have a large enough supply of electricity to meet demand. In order to deal with the shortage, utility companies shut off power to certain neighborhoods for several specified hours. In the case of the 2000 rolling blackouts in California, the problem was not one of supply. California had enough generating capacity to provide electricity for its residents. The problem arose with large energy providers that manipulated the energy markets to enhance profits. The most well-known example of this practice was the abuses of Enron Corporation. The company used energy trading schemes to increase its profits as the cost of energy for consumers skyrocketed. In one example, California's consumers paid in excess of $5.5 million for their electricity while Enron recorded a $10 million profit—in one day (Swartz and Watkins 2003, 240)!

The case of California demonstrates why regulation is necessary in a market prone to monopoly. Despite abuses like these, many still contend that deregulation would provide a more responsive and adaptive energy system for consumers. This view was expressed in the Energy Policy Act of 2005. While the legislation provided protection to utility customers from Enron-like scandals, it also repealed the 1935 PUHCA. The repeal of PUHCA is significant because it rescinded the fundamental legislation that dissolved the corrupt activities of utility holding companies in the 1930s. The 2005 act also repealed the PURPA requirement that utilities must purchase electricity from qualifying facilities (UCS 2005).

Conclusion

This chapter described energy dynamics of the United States. This country is the most energy-intensive society in the world. Throughout history, it has gone to great lengths to secure energy supplies. Despite these efforts, a comprehensive energy policy has remained elusive. This lack of a unified policy is due to the complex nature of energy use, as well as to political tensions and special interests that have steered the direction of energy and environmental policy throughout U.S. history. This history has been complex and multifaceted. In an effort to summarize information on energy use history and issues, the next chapter provides a chronological overview of the first three chapters.

References

Clarke, J. N., and K. Angersbach. 2001. "The Federal Four: Change and Continuity in the Bureau of Land Management, Fish and Wildlife Service, Forest Service and National Park Service, 1970–2000." In *Western Public Lands and Environmental Politics*, edited by C. Davis. 2nd ed. Boulder, CO: Westview Press.

Cronon, W. 2001. "Neither Barren nor Remote." *New York Times*, February 28, Section A, p. 19.

Davis, D. H. 2001. "Energy on Federal Lands." In *Western Public Lands and Environmental Politics,* edited by C. Davis. 2nd ed. Boulder, CO: Westview Press.

Energy Information Administration (EIA). 2006a. "Table 1.3: Energy Consumption by Source." *Monthly Energy Review* (January 2006): 7.

Energy Information Administration (EIA). 2006b. "U.S. Crude Oil and Petroleum Products Exports." http://tonto.eia.doe.gov/dnav/pet/hist/mttexus1A.htm (accessed January 10, 2006).

Energy Information Administration (EIA). 2006c. "U.S. Crude Oil and Petroleum Products Imports from All Countries." http://tonto.eia.doe.gov/dnav/pet/hist/mttimus1A.htm (accessed January 10, 2006).

Energy Information Administration (EIA) 2005a. *Annual Energy Review 2004.* Washington, DC.: DOE/EIA-0384(2004).

Energy Information Administration (EIA). 2005b. "Table 14. Recoverable Coal Reserves and Average Recovery Percentage at Producing Mines by State, 2004, 2003." *Annual Coal Report, 2004.* DOE/EIA-0584(2004). September. Washington, DC: Department of Energy/Energy Information Administration.

Energy Information Administration (EIA). 2005c. "World Proved Reserves of Oil and Natural Gas, Most Recent Estimates." http://www.eia.doe.gov/ (accessed January 10, 2006).

Fehner, T. R., and J. M. Holl. 1994. *Department of Energy 1977–1994: A Summary History.* DOE/HR-0098.

Gorman, H. S. 2001. *Redefining Efficiency: Pollution Concerns, Regulatory Mechanisms, and Technological Change in the U.S. Petroleum Industry.* Akron, OH: University of Akron Press.

Idelson, H. 1992. "National Energy Strategy Provisions." *Congressional Quarterly Weekly Report* 50 (47): 3722–3731.

Kahn, J. 2001a. "Cheney Promotes Increasing Supply as Energy Policy." *New York Times,* May 1, Section A, page 1.

Kahn, J. 2001b. "Cheney Withholds List of Those Who Spoke to Energy Panel." *New York Times,* June 26, Section A, p. 17.

Kapstein, E. B. 1990. *The Insecure Alliance: Energy Crises and Western Politics Since 1944.* New York: Oxford University Press.

Laitos, J. G., and J. P. Tomain. 1992. *Energy and Natural Resources Law in a Nutshell.* St. Paul, MN: West Publishing.

Long, J. C. S., and R. C. Ewing. 2004. "Yucca Mountain: Earth-Science Issues at a Geologic Repository for High-level Nuclear Waste." *Annual Review of Earth & Planetary Sciences* 32 (1): 363–401.

Melosi, M. V. 1985. *Coping with Abundance: Energy and Environment in Industrial America.* Philadelphia, PA: Temple University Press.

Miller, E. W., and R. M. Miller. 1993. *Energy and American Society: A Reference Handbook.* Santa Barbara, CA: ABC-CLIO.

Minerals Management Service (MMS). 2003. "MMS Facts and Figures 2003." http://www.mms.gov/ooc/newweb/publications/2003%20FACT.pdf (accessed February 6, 2006).

Nivola, P. S. 1986. *The Politics of Energy Conservation.* Washington, DC: The Brookings Institute.

Nuclear Energy Institute Inc. v. Environmental Protection Agency. 373 F.3d 1251 (D.C. Cir. 2004).

Office of Civilian Radioactive Waste Management (OCRWM). 2006. "Why Yucca Mountain?" http://www.ocrwm.doe.gov/ymp/about/why.shtml (accessed February 7, 2006).

Reblitz-Richardson, B. 2005. "D.C. Circuit Rejects EPA's Proposed Standards and Extends Timeline for Yucca Mountain Nuclear Waste Repository." *Ecology Law Quarterly* 32 (3): 743–748.

Revkin, A. C. 2001. "Hunting for Oil: New Precision, Less Pollution." *New York Times,* January 30, Section F, p. 1.

Rosenbaum, W. A. 1973. *The Politics of Environmental Concern.* New York: Praeger.

Seelye, K. Q. 2001. "Bush Promotes Energy Bill as Security Issue." *New York Times,* October 12, Section A, p. 18.

Seelye, K. Q. 2002. "Bush Favors Dozens of Sites for Exploration." *New York Times,* April 19, Section A, p. 18.

Smith, Zachary A. 2004. *The Environmental Policy Paradox.* 4th ed. Upper Saddle River, NJ: Prentice Hall.

Streever, B. 2002. "Science and Emotion on Ice: The Role of Science on Alaska's North Slope" *BioScience* 52 (2): 79–84.

Sullivan, C. C. 1993. "Economy and Ecology: A Powerful Coalition." *Buildings* 87 (5): 47.

Swartz, M., and S. Watkins. 2003. *Power Failure: The Inside Story of the Collapse of Enron.* New York: Doubleday.

Union of Concerned Scientists (UCS). 2005. "Summary of the Energy Bill: Conference Report to HR 6." http://www.ucsusa.org/clean_energy/clean_energy_policies/energy-bill–2005.html (accessed February 6, 2006).

U.S. Department of Energy (DOE). 1998. *Comprehensive National Energy Strategy.* Washington, DC: DOE/S-0124.

U.S. Environmental Protection Agency (EPA). 2005. "Clear Skies: Basic Information." http://www.epa.gov/air/clearskies/basic.html (accessed February 6, 2006).

U.S. Geological Survey (USGS). 1998. "Arctic National Wildlife Refuge, 1002 Area, Petroleum Assessment, 1998, Including Economic Analysis." http://pubs.usgs.gov/fs/fs-0028-01/fs-0028-01.htm (accessed January 21, 2006).

Van Natta, D., Jr., and N. Banerjee. 2002. "Top G.O.P. Donors in Energy Industry Met Cheney Panel." *New York Times*, March 1, Section A, p. 1.

Verhovek, S. H. 2002. "Drilling Could Hurt Wildlife, Federal Study of Arctic Says." *New York Times*, March 30, Section A, p. 12.

4

Chronology

Introduction

U nderstanding the current state of energy dynamics in society would be impossible without consideration of historical trends. Such hindsight also provides a valuable component in the development of sustainable energy policy. In this chapter, chronologies summarize and highlight important events in global energy use. The purpose of this chapter is to provide a useable format for providing quick reference to important energy events.

Energy dynamics permeate many areas of human life. Listing associated events in one extended timeline does not fully highlight their relevance in society. Additionally, such a format may be frustrating to readers seeking information relevant to one aspect of energy use. For this reason, energy events are divided into six chronologies within this chapter. The first four describe notable milestones and achievements in the development and production of fossil fuels, nuclear energy, renewable resources, and energy services such as electricity and transportation. The fifth chronology details important political and economic events in global energy history, while the sixth focuses on U.S. events.

Fossil Fuels: Coal, Petroleum, and Natural Gas

Carboniferous Period (286 million to 360 million years ago)
Bituminous coal found in the eastern United States and Europe is formed.

Permian Period (240 million to 286 million years ago)
Coal found in eastern Asia, Siberia, western United States, Indonesia, and Australia is formed.

2000 BCE Ancient Egyptians use petroleum oil for medicinal purposes.

1000 BCE China uses coal for smelting copper.

480 BCE Persians use oil as a flammable material in warfare during invasion of Athens.

200 BCE China uses percussion drilling to extract natural gas for salt-brine evaporation.

1200s CE Coal is extensively mined in Europe for metal smelting.

1709 Abraham Darby develops a technique for producing pig iron using coke (pyrolyzed coal).

1748 The first commercial coal production in the United States begins near Richmond, Virginia, where coal was discovered in 1701.

1800 Great Britain is the largest global coal producer, supplying over four-fifths of global coal resources.

1807 The first municipal coal-gas system lights up Pall Mall in London.

1816 Baltimore, Maryland, is the first city in the United States to use coal gas for lighting.

1821 William Hart digs the first natural gas well in the

United States in Fredonia, New York. He later founds the Fredonia Gas Light Company.

1830s   Coal companies flourish in the Appalachian regions in the United States and along the Ohio, Illinois, and Mississippi rivers.

1850s   Coke is the dominant fuel used in English blast furnaces.

1853   Abraham Gesner separates kerosene from gasoline using a process called distillation.

1859   Colonel Edwin Drake makes the first oil strike in the United States in Titusville, Pennsylvania.

1866   Strip mining for coal in the United States begins near Danville, Illinois.

1880s   Petroleum production begins in Russia with the development of the Baku oil fields.

1890s   Royal Dutch, an Indonesian company, begins oil extraction in the Dutch East Indies.

1891   The first natural gas pipeline is constructed in the United States, extending 120 miles and supplying Chicago with natural gas from gas fields in Indiana.

1901   Foreign oil companies begin production in Mexico.

        The technique of rotary drilling is used for the first time to develop the Spindletop wells in Beaumont, Texas.

1908   William D'Arcy strikes oil at Masjid-i Suleiman in Persia (modern-day Iran).

1913   Foreign oil companies begin production in Trinidad.

1914   Foreign oil companies begin production in Venezuela.

1925 Annual global petroleum production exceeds 1 billion barrels.

1930s The United States begins constructing natural gas pipeline networks, most notably completing a pipeline that extends from western Texas to Chicago.

1938 The California Standard Oil Company (CASOC) strikes oil in Saudi Arabia.

1940s The United States expands its gas pipeline network.

1940 Annual global petroleum production exceeds 2 billion barrels.

1960 Oil production begins in the oil fields of Daqing, China's largest oil fields.

   Oil is discovered in the Konda River valley in western Siberia, marking the beginning of oil production in Siberia.

1970s The world's largest oil pipelines are built to transport petroleum from Siberia to Europe.

1980s Conflict between Iran and Iraq cause unstable fluctuations in the global oil market.

1996 Shell Oil announces plans to drill for oil at record depths of 4,000 feet in the Gulf of Mexico. Production tests began in 2006.

1999 British Petroleum Company (U.K.) and Aramco (fifth-largest U.S. oil company) sign a $53 billion merger.

2000 A large natural gas reserve is discovered in the Tarim Basin in the Xinjiang region of Western China.

2001 The first stage of the Caspian Pipeline, a 1,510-kilometer-long oil pipeline from the Tengiz field in western Kazakhstan to Russia's Black Sea coast opens with the capacity to transport 350,000 barrels of oil per day.

2005   Royal Dutch Shell reduces it estimates of energy reserves by 1.4 billion barrels of oil equivalent.

Oil giant Chevron (U.S.) purchases U.S.-based Unocal for $16.7 billion increasing its oil reserves by 1.8 billion barrels of oil equivalent.

Chevron begins construction on the West African Gas Pipeline, which will stretch 450 miles from Nigeria to Benin, Ghana, and Togo.

2006   A 600-mile oil pipeline stretching from Kazakhstan to China is completed with a capacity of delivering 210,000 barrels per day.

Nuclear Energy

1896   French physicist Antoine-Henri Becquerel discovers the radioactive properties of uranium.

1898   French scientists Pierre and Marie Curie discover the radioactive elements plutonium and radium.

1908   British physicists Ernest Rutherford and Frederick Soddy discover alpha and beta radiation and describe the theory of radioactive transformation.

1938   German physicists Otto Hahn and Fritz Strassmann demonstrate the phenomenon of nuclear fission.

1942   Enrico Fermi sustains the first controlled nuclear chain reaction for 28 minutes at the University of Chicago.

1945   The first atomic bomb is denoted on July 16 near Alamogordo, New Mexico.

The United States drops an atomic bomb on the Japanese city of Hiroshima on August 6. On August 9, the United States detonates a second atomic bomb on Nagasaki, Japan.

1951   The first breeder reactor produces useable electric power from atomic energy, illuminating four lightbulbs.

1952   The first hydrogen bomb, a device 1,000 times more powerful than the atomic bomb, is detonated on the Pacific Island of Eniwetok Atoll.

1954   The first nuclear power plant commences operation in Obninsk, Russia, marking the first time that electricity is derived from nuclear energy for civilian use.

1955   Arco, Idaho, is the first U.S. town to receive electricity generated using nuclear energy; the power comes from the Idaho National Energy Laboratory, a U.S. Department of Energy facility.

1957   The first large-scale nuclear power plant, the Shippingport Atomic Power Station, commences operation in Pennsylvania.

    An English nuclear power plant, Windscale Pile Number One, catches on fire, releasing radioactive contaminants into the atmosphere.

1958   A moratorium is placed on nuclear weapons testing operations.

1961   The United States and the Soviet Union resume nuclear weapons testing.

    An experimental nuclear reactor explodes at the Idaho National Reactor Testing Station, killing three people.

1962   The first nuclear power plant in Antarctica commences operation.

1963   The Limited Tests Ban Treaty is signed by the United States, Russia, and Great Britain, banning the underwater, atmospheric, and outer-space testing of nuclear weapons.

1964 Private ownership of nuclear fuel in the United States is allowed through the Private Ownership of Nuclear Materials Act.

1968 The Nuclear Nonproliferation Treaty is signed by forty-eight countries. Aimed at halting the spread of nuclear weapons, the treaty details provisions for the peaceful development of nuclear technology. The treaty is fully ratified in 1970.

1977 A fire at the Brown's Ferry Nuclear Power Plant in Alabama causes a malfunction of safety systems at the facility. Although no radioactive material was released to the environment, the accident raises concerns about nuclear safety.

1979 Loss of coolant at the Three Mile Island nuclear power plant in Pennsylvania causes an accident that raises public fears about the nuclear power industry.

1986 A nuclear reactor meltdown occurs in at Chernobyl, in Soviet-controlled Ukraine. The accident kills thirty-one people and spreads radiation over Europe.

2000 The Nuclear Regulatory Commission grants twenty-year operating extensions to two U.S. nuclear power plants.

2002 The world's first nuclear power plant shutdown happens in Obninsk, Russia.

2004 Great Britain closes its Chapelcross nuclear power plant.

   Citing safety concerns, Lithuania implements plans to shut down one-half of its nuclear generating capacity.

2005 Germany closes its Obrigheim nuclear power plant as part of an initiative to close all seventeen nuclear power facilities in the country by 2021.

2005 *(cont.)*	China experiences nationwide power shortages as electric capacity struggles to meet rapidly increasing demand in response to record high temperatures.
2006	Turkey announces plans for the construction of its first nuclear power plant, located at Sinop.

Renewable Energy

2 BCE	Waterpower is used to mill grain in Middle Eastern and Scandinavian countries.
1100s CE	Wind power is used in Europe for the purpose of milling grain.
1600s	Waterpower is the main source of energy for milling grain in Europe. By the end of this century, England has more than 20,000 water mills.
1833	Benoit Fourneyron develops the first water turbine, an invention that revolutionizes the use of water for powering mills and other industrial machinery. Fourneyron's design converts potential energy stored in water to useful mechanical energy with 80 percent efficiency.
1839	French physicist Edmond Becquerel discovers the photovoltaic effect when he measures an increase in voltage of a battery exposed to sunlight.
1877	British scientists William Grylls Adams and Richard Evans Day discover that the element selenium exhibits electrical properties when it is exposed to sunlight. This observation leads to the use of selenium in photovoltaic solar cells.
1882	Water turbines are coupled to electricity generators for the first time in the United States.
1883	Charles Edgar Fritts of New York develops the first

selenium solar cell. The electricity conversion in the cell is only around 1 percent.

1888   Charles F. Brush uses wind to create power in Cleveland, Ohio, marking the first time that a large windmill in used to generate electricity in the United States.

1904   Italy is the first country to use geothermal energy to produce electricity.

1909   William J. Bailey patents the solar water heater.

1920   Southern Florida initiates development of the solar heater market. Although business stagnates by the 1950s, the effort marks the first time that solar energy is shown to be commercially viable in the United States.

1941   Researchers at Bell Laboratories in the United States discover photovoltaic properties of the element silicon.

1954   American researchers Daryl Chapin, Calvin Fuller, and Gerald Pearson develop a silicon-based photovoltaic solar cell with an energy conversion efficiency of 6 percent.

1958   Photovoltaic solar cells are used on the space satellite Vanguard I to power a small radio transmitter.

1960   Pacific Gas & Electric starts up the first commercially viable 10-megawatt geothermal generating station in the United States.

1961   Construction begins on the world's first tidal power plant on the Rance River estuary in Brittany, France; the plant first generates power in 1966.

1968   The Soviet Union opens a tidal power station in Murmansk.

1974	The first two-bladed wind turbine is developed by the National Aeronautics and Space Administration (NASA), marking a shift in wind turbine technology. This design was perfected throughout the 1970s.
	The Solar Energy Industries Association (SEIA) is formed to provide lobbying support for the solar industry in Washington, DC.
1977	The Solar Energy Research Institute creates the first federally funded research and development lab for renewable energy.
	The first hot dry-rock reservoir is developed to exploit geothermal energy in Fenton Hill, New Mexico.
1980s	California provides generous tax cuts for wind energy development; wind farms are developed throughout the state.
1980	The Solar Rating and Certification Corporation is established for the purpose of developing standards for solar equipment.
	The first U.S. power plant comprised of photovoltaic solar cells opens in Utah as an experimental generating station.
1983	The first solar electric generating station, named SEGS-I, is installed in Southern California.
1984	The first tidal power plant in North America commences operation in Canada.
1986	The Itaipu Dam commences operation along the Brazil and Paraguay border. At the time, it is the largest dam in the world.
1988	The first European photovoltaic power station is built near the city of Koblenz, Germany, for the purpose of testing the feasibility of photovoltaic power contribution to local and regional power grids.

1990   General Motors introduces the first electric vehicle.

1991   Approximately 15,500 wind turbines are in operation in California, producing 2,700 MkWh of electricity.

The first wind farm in the United Kingdom commences operation in Cornwall.

1993   U.S. Windpower makes commercially available a variable speed wind turbine.

1994   There is an estimated 3.5 GW of installed wind power capacity worldwide.

Electricity, Engines, Lights, and Energy Services

1690   Denis Papin builds the first small coal-powered steam engine. Thomas Savery and Thomas Newcomen later modify the design. By 1750, water-pumping steam engines are installed in English mines.

1765   James Watt expands the steam engine design by adding a separate condenser, thereby increasing efficiency and power output. Watt's innovations mark the rise in size and use of the modern steam engine (see Watt's biography in chapter 5).

1802   The *Charlotte Dundras*, built by Patrick Miller in England, is the first commercially successful ship to be powered by a steam engine.

1808   Sir Humphrey Davy develops arc lighting, the first form of electric lighting, in England.

1821   Michael Faraday discovers the principle of electromagnetic induction, the basic physical principle that explains how electricity is generated by magnetism.

1830   Steam engines become the primary energy sources used for land and water transport. Steam-powered locomotives and ships create the possibility for global transportation and shipping networks.

The first public railway, powered by steam locomotives, operates between Liverpool and Manchester in England.

1833   The *Royal William* is the first steam-powered ship to cross the Atlantic Ocean, traveling from London to Quebec.

1856   Steel is made using blast furnace technology, a process developed independently by Henry Bessemer and William Kelly.

1868   The open hearth method for making steel, known as the Siemens-Martin process, is developed.

1869   The first transcontinental railway link is completed in the United States.

1878   Nikolaus Otto develops and patents a coal-powered four-stroke horizontal internal combustion engine.

1879   Thomas Edison invents the first lightbulb by enclosing a carbonized sewing thread inside a glass under vacuum.

1882   Thomas Edison develops the first commercial electric system using direct current (DC) at 110 volts for electricity transmission.

America's first power plant, the Pearl Street Station, is commissioned in New York.

Lewis Latimer invents an inexpensive method of manufacturing carbon filaments for electric lightbulbs. This work sets the foundation for widespread lighting capabilities.

The first U.S. hydroelectric station opens in Wisconsin.

1884   Charles Parsons introduces the first steam turbine in England. His invention is a smaller, more efficient, more powerful alternative to Watt's steam engines.

1885   George Westinghouse develops alternating current (AC) electrical systems.

William Stanley invents the transformer, a device that allows for efficient electricity transmission and delivery.

Robert Bunsen invents the Bunsen burner, a device that uses natural gas to create a flame for cooking and heating.

Karl Benz builds the first car powered by a horizontal gasoline engine.

Gottleib Daimler and Wilhelm Myabach invent a high-speed single-cylinder vertical gasoline engine.

1888   Nikola Tesla patents the three-phase AC electric motor.

1892   Rudolf Diesel patents the first diesel-powered engine. This invention revolutionizes land and water transport, replacing steam engines by the mid-twentieth century.

1896   Henry Ford builds his first car.

1903   In North Carolina, Wright brothers Orville and Wilbur achieve the first airplane flight.

1911   Electric air conditioning is first used.

1913   The electric refrigerator is invented.

1917   The first Russian transcontinental rail link is completed, extending from Siberia to Vladivostok.

1928	The construction of Boulder Dam in Arizona is the first large dam constructed in the West that incorporates hydropower in its principal design.
1930s	Aviation was revolutionized when Frank Whittle and Hans Pabst build the first experimental gas turbines for powering military aircraft.
1958	The Boeing 707 is introduced as the first commercial passenger airplane.
1998	The electric utility company Detroit Edison receives funding from the U.S. Department of Energy (DOE) to construct the first high-temperature superconductor power cable, a technology designed to increase the reliability and capacity of the electricity grid.
2004	The first U.S. hydrogen refueling station opens in Washington, DC.
2006	The first wave power plant becomes operational off the coast of Portugal.

World Energy

1884	The Anglo-American Company is formed as Standard Oil Company's foreign associate.
1906	Shell Transport and Trading Company merges with Netherlands oil company Royal Dutch to form one of the largest global oil producers.
1909	William D'Arcy founds the Anglo-Persian Company. The company changes its name to British Petroleum (BP) in 1914 when the British government purchases half of its holdings.
1912	The Turkish Petroleum Company (TPC) is established in Iraq.

1914   World War I begins in Europe. The United States plays a vital role in supplying the Allied Powers with energy resources.

1922   The TPC allows the entry of U.S. companies into its holdings, marking the first time U.S. companies are allowed to develop oil in the Middle East.

1928   The Red Line Agreement is signed, forming an alliance among Iran Petroleum Company (IPC, previously the TPC) companies that outlines concessions to drill in the Middle East.

The Achnacarry ("As-Is") Agreement is signed, creating a secret arrangement among major oil companies to fix prices to the Gulf-Plus System.

1933   King Saud, the ruler of Saudi Arabia, grants oil exploration concessions to the U.S. oil giant Socal, marking the entry of U.S. energy companies in the Middle East.

1936   The Arabian-American Oil Company (ARAMCO) is formed when Texaco joins Socal's concessions in Saudi Arabia.

1937   World War II begins in Asia, increasing the demand for energy production.

1938   The Mexican expropriation occurs when Mexico nationalizes its petroleum industry, forming the state-owned and -operated company, Pemex. This marks the first time that a developing country seizes control of its oil resources from foreign ownership.

1945   The European Coal Organization (ECO) is formed, making it the first transnational alliance to respond to an energy crisis.

1946   The United Nations Atomic Energy Commission (UNACE) is created to promote peaceful nuclear development.

1947   The United States approves of the Marshall Plan pro-
            viding economic and energy aid to Europe during the
            postwar reconstruction period.

            The ECO is dissolved, and the European Coal and
            Steel Community (ECSC) is formed in its place.

1948   The Red Line Agreement is dissolved.

            The Fifty-Fifty Agreement is signed by Venezuela and
            oil companies operating within the country, dividing
            all profits equally.

1950   The Trans-Arabian oil pipeline (Tapline) is completed,
            allowing Saudi Arabian oil to be delivered to the
            Mediterranean Sea.

1951   Mohammed Mossadegh rises to power in Iran, mark-
            ing the beginning of the Iran crisis. Opposing BP con-
            cessions in the country, he attempts to nationalize
            Iran's oil industry.

1953   Mohammed Mossadegh is assassinated in a coup led
            by the U.S. Central Intelligence Agency (CIA), mark-
            ing the first time that a clear link is made between U.S.
            foreign policy and energy policy.

1954   Seventy-one countries sign the International Conven-
            tion for the Prevention of Pollution of the Sea by Oil.
            This treaty is the first international attempt to reform
            oil tanker practices of hull flushing and oil dumping
            into the sea.

1955   The ECSC forms EURATOM to promote nuclear de-
            velopment and technology in European nations.

1956   The Suez Crisis closes the Suez Canal to oil shipments,
            causing a massive energy crisis in Europe.

            The International Atomic Energy Agency is formed
            with eighty-one member countries.

1957   The Organization for European Economic Cooperation (OEEC) is established is response to the Suez Crisis.

1960   The Organization of Petroleum Exporting Countries (OPEC) is formed.

1961   The Organization for Economic Cooperation and Development (OECD) is formed in response to OPEC. This organization replaces the OEEC and extends membership to the United States, Canada, Japan, New Zealand, and Australia.

1967   Israel preemptively attacks Egypt, marking the beginning of the Six Day War. Arab States implement an oil embargo against the United States and Europe for their support of Israel. Although the embargo was lifted after the war, tensions between Arab and non-Arab countries remain high.

               The vessel *Torrey Canyon* releases over 36 million gallons of crude oil into the English Channel.

1968   Arab states form the Organization for oil-producing Arab Petroleum Exporting Countries (OAPEC) to unite the political interests of Arab nations.

1971   OPEC's Tehran Agreement results in an eighty-cent tax increase per barrel by 1975.

1973   The Yom Kippur War, also known as the Arab-Israeli War, begins. U.S. support for Israel during this war leads to an oil embargo against the United States, resulting in the most crippling energy crisis in American history.

               The International Convention for the Prevention of Pollution from Ships (MARPOL) is signed by 125 countries. It mandates that oil tankers and receiving ports revise their practices to reduce oil pollution into the sea when transporting petroleum.

1974   The International Energy Agency (IEA) is formed.

1977   Nigeria creates the Nigerian National Petroleum Corporation (NNPC) to regulate Nigeria's oil industry and joint venture contracts with foreign oil companies.

1978   The Iranian Revolution begins, resulting in a drop in oil production. Reduced production continues throughout the early 1980s as security issues intensify in the Middle East and OPEC requires low production to keep oil prices high.

            China develops its modernization plan, a strategy that stipulates quadrupling the country's industrial and agricultural output by the year 2000.

            Breakup of the *Amoco Cadiz* oil tanker releases 65 million gallons of oil into the ocean off the coast of France.

1979   The oil tanker *Atlantic Empress* spills 76 million gallons of crude oil into the ocean off the coast of the West Indies.

            A blowout on an oil platform, IXTOC 1 well, releases 140 million gallons of oil into the Gulf of Mexico.

            The Convention on Long-Range Transboundary Air Pollution is signed by forty-nine countries. The treaty went into force in 1983.

            China signs contracts with sixteen foreign oil companies to conduct geophysical surveys of its energy resources.

1980   The Iran-Iraq War begins reducing oil production in the Middle East.

1982   China establishes the China National Offshore Oil Corporation to oversee drilling contracts with foreign entities.

1986   Saudi Arabia removes petroleum production controls, resulting in a surplus of oil on the global market. Petroleum prices drop, stimulating overconsumption.

1989   The dissolution of the Soviet Union results in a large decrease in global energy demand and output from a significant energy-producing region.

1990   Iraq invades Kuwait on August 2 resulting in a petroleum price increase.

1991   Iraqi armed forces deliberately damage oil pipelines flowing to ports in Kuwait as an offensive measure during the Gulf War. The leaks release an estimated 240 million gallons of oil into the ocean.

1992   The United Nations Framework Convention on Climate Change is signed by 188 countries, marking the first international recognition of the issue of climate change. The treaty went into force in 1994.

1997   The Asian Economic Crisis begins in Thailand. The collapse of a large number of Asian economies decreases world energy demand.

                The Kyoto Protocol on Climate Change is signed by 120 countries. The treaty went into force in 2004 when Russia ratified the agreement.

1998   Global oil prices drop significantly as a result of the Asian Economic Crisis and OPEC's high production quotas.

                India and Pakistan begin testing nuclear weapons.

2000s   In Nigeria, attacks against employees of foreign-owned oil companies and oil production infrastructure increase, temporarily halting production in some areas of the Niger Delta.

2003   The United States invades Iraq in March, resulting in a drop in oil production from Iraq.

2004   Nigeria's government issues a mandate requiring
            Royal Dutch Shell to pay $1.5 billion to the Ijaw
            people for environmental and health damage caused
            from oil operations.

2005   The European Union (EU) opens the first greenhouse
            gas (GHG) emissions trading system in Europe allow-
            ing GHG emitters the ability to trade carbon credits.

            The Kyoto Protocol goes into effect.

2006   Two hundred people in May and 269 people in De-
            cember are killed in Nigeria after explosions occur
            along oil pipelines that have been damaged to extract
            oil for sale on the black market.

U.S. Energy

1813   A fuel crisis occurs in Philadelphia as coal prices in-
            crease from $.30 a bushel to $1.00 in one month.

1840s   Wood is the primary energy resource used in America
            for domestic and industrial purposes.

1854   The New York Kerosene Company is founded by
            Abraham Gesner. It is the first company in the United
            States to manufacture and distribute coal oil for illu-
            mination.

1870   The consumption of wood as a primary energy source
            peaks.

            John D. Rockefeller establishes the Standard Oil
            Company.

1882   The Standard Oil Trust is established.

1890s   Antismoke coalitions form throughout the country to
            raise public awareness about the negative health im-
            pacts of smoke pollution in urban areas.

1892   The General Electric Company (GE) is formed by a merger of the Edison Company and Thomas-Houston.

Edward L. Doheny discovers oil near Los Angeles, California.

1901   An extensive oil field is discovered in Spindletop, Texas, giving rise to the Texas Company (Texaco) and Gulf Oil, two dominant global oil companies of the twentieth century.

1908   Henry Ford ushers in mass production when he develops rapid assembly lines to cost-effectively produce his Model T. The cheap vehicles create demand for gasoline.

1910   The U.S. Bureau of Mines is established to develop and enforce coal mine safety standards.

1913   William Burton patents a catalytic cracking technique that converts oil to gasoline.

1917   The United States enters World War I. The federal government establishes the U.S. Fuel Administration and the National Petroleum War Services Committee to help with wartime production, allocation, and distribution of energy resources.

A fuel shortage stimulates an energy crisis in Eastern cities.

1918   Crude oil flows through the first U.S. pipeline in Wyoming.

1920s   Coal ceases to be the dominant energy source in the United States as the production and consumption of oil increases.

1920   Nine million automobiles are operating in the United States.

1930s   New Deal policies are enacted, creating an expansion of publicly funded power projects and promoting the development of electricity in rural areas.

1941   The Office of Petroleum Coordination for National Defense is established for the purpose of controlling petroleum production and consumption during World War II.

1942   The War Production Board is established for the purpose of coordinating industry for war production.

1943   Two major oil pipelines, "The Big Inch" and the "Little Big Inch," are completed to deliver petroleum to the East Coast.

1946   The Oil and Gas Division (OGD) is established by the federal government to develop a database on oil and gas demand and to coordinate the implementation of oil and gas policy.

         The National Petroleum Council, comprised of industry executives, is established to serve as an advisory board to the federal government.

1947   The United States becomes a net importer (rather than a net exporter) of oil.

1953   President Dwight Eisenhower delivers the famous "Atoms for Peace" speech before the United Nations delegation in which he calls for the establishment of an International Atomic Energy Agency to oversee the development and use of fissionable materials worldwide.

1959   President Dwight Eisenhower establishes the Mandatory Oil Import Program (MOIP) limiting petroleum imports to a specified amount as a way to stabilize the domestic oil market.

1960s   Domestic electricity consumption increases with the widespread distribution of television sets.

1965   A switch malfunction at a power plant in Ontario causes a massive blackout in the eastern United States and parts of Canada, shutting power off for 13 hours to an estimated 30 million people.

1968   Oil is discovered in Prudhoe Bay along the North Shore of Alaska.

1969   The National Environmental Policy Act (NEPA) is enacted mandating that an Environmental Impact Statement be drafted for development or extractive projects on federal land.

         A blowout on an oil platform off the coast of Santa Barbara, California, releases 230,000 gallons of crude oil into the ocean, polluting beaches and the California Pacific Coast.

1970   Electricity brownouts occur in the Northeast as a result of an increase in electricity demand during a heat wave.

         The Environmental Protection Agency (EPA) is established.

1972   Oil production reaches its highest level in the United States.

1973   The Arab Oil Embargo stops shipments of oil from Arab producing countries to the United States, resulting in an energy crisis. President Richard Nixon establishes Project Independence in response to the Arab Oil Embargo. This plan aims to create energy self-sufficiency in the United States. The Federal Energy Office is created to oversee fuel pricing structures, oil rationing programs, and pricing.

1974   Construction begins on the Trans-Alaska Pipeline System (TAPS).

         The American Wind Energy Association is founded.

1977   President Jimmy Carter announces the National Energy Plan (NEP) on April 18. The plan calls for energy conservation measures, implementation of the Crude Oil and Equalization Tax (COET), and the integration of inter- and intrastate natural gas markets.

             Oil is pumped through the TAPS for the first time.

             The United States commences storage of the Strategic Petroleum Reserves in Louisiana.

1978   Important energy legislation is passed, including the Public Utilities Regulatory Policies Act (PURPA) and the Natural Gas Act.

1981   President Ronald Reagan removes price and allocation controls on the oil industry.

             President Reagan lifts the ban on reprocessing of nuclear fuel in an effort to stimulate the nuclear power industry.

1984   The National Coal Council is established to act as an industrial advisory committee to the federal government.

1987   Yucca Mountain is selected as the United States' national nuclear waste repository.

1988   President Reagan repeals the Windfall Profits Tax that is levied against the profits of oil companies.

1989   The *Exxon Valdez* oil spill releases 11 million gallons of crude oil into Alaska's Prince William Sound, creating an oil slick over 900 square miles.

1993   The Climate Change Action Plan initiated by President Bill Clinton and Vice President Al Gore calls for voluntary measures to reduce greenhouse gas emissions to 1990 levels by the year 2000.

1996   The United States closes its largest plutonium

processing plant, the Plutonium Uranium Extraction Facility, used throughout the Cold War.

California governor Pete Wilson signs a law restructuring California's energy markets.

2000   Rolling blackouts occur in California as a result of market restructuring.

2003   On August 14, the largest blackout in U.S. history leaves most of the northeastern United States and parts of Canada without power for several days.

Oil companies ConocoPhilips and Anadarko Petroleum Corp. are authorized to develop the National Petroleum Reserve around the Alpine Field along the North Slope of Alaska, marking the first time that national reserves are approved for development.

2005   Hurricane Katrina hits the city of New Orleans in Louisiana, temporarily halting oil and natural gas production in the Gulf of Mexico and shutting down oil refineries in the southern United States, causing gasoline prices to skyrocket.

The United States government agrees to release 30 million barrels of crude oil from the Strategic Petroleum Reserve.

2006   In August, British Petroleum Company shuts down production along the eastern half of its production fields along the North Slope of Alaska after discovering multiple leaks in the pipeline network. Production resumes in September.

Transport of crude oil and natural gas along the High Island Pipeline System (HIPS) in the Gulf of Mexico is shut down after it is damaged from a ship anchor.

5

Biographical Sketches

Introduction

The biographies in this chapter are glimpses into the roles of people who had a large impact on the development, distribution, and use of energy resources. Because energy in society is a complex and multifaceted topic, the people who have influenced energy dynamics in modern society include a diverse array of characters. Some sketches portray the lives of inventors and scientists notable for their contributions to the development of energy technology and the understanding of energy's physical properties. Other sketches examine political leaders important for their roles in the development and promotion of energy resources in their respective countries, and energy business and economic leaders whose work has contributed to the modern structure of energy markets. Also included are biographies of leaders in the environmental and social movements who contributed to energy dynamics by raising awareness about the negative consequences of energy use.

It is important to note that this selection of biographies profiles only a few of the many notable people who have contributed to energy dynamics in society. As stated in previous chapters, energy is an important issue in many aspects of society. It would be impossible to include an exhaustive list of individuals in one chapter. The key figures included were chosen because they further illustrate the complexity of energy use in society and provide the reader with a holistic understanding of energy issues.

Juan Perez Alfonzo (1903–1979)

Born in Caracas, Venezuela, Juan Perez Alfonzo was one of the most influential political figures involved in global energy dynamics in the twentieth century. Holding bachelor's degrees in physics and mathematics and a doctorate in political and social sciences, he became involved in politics in 1936 when he joined the Venezuelan Organization (Organización Venezolana, or ORVE) and became a member of the National Democratic Party in Venezuela. He became the minister of promotion for the Venezuelan government in 1945. During that time, he crafted the famous Fifty–Fifty Agreement between oil companies and the Venezuelan government, a policy reform that sparked a trend toward energy nationalization in Latin American countries. Alfonzo's most notable influence in global energy dynamics occurred during his tenure as the minister of mines and hydrocarbons. Holding this title, he served as the delegate to the Arab Oil Congress in 1959, where he suggested the development of an "oil consultation commission" to regulate oil production in producing countries. This idea was the foundation for the Organization of Petroleum Exporting Countries (OPEC), which enabled large oil-producing countries to influence production and pricing in the global oil market. In 1961, Alfonzo published the book *Petroleum: Earth Juice*. He retired from his position as minister of mines in 1963. He died of cancer in 1979 in the United States when he was 76 years old.

Mahmoud Ahmadinejad (1956–present)

Elected as president of Iran in 2005, Ahmadinejad is an important contemporary figure in global energy politics. Born in Garmsar, Iran, in 1956, his family soon relocated to Tehran. In 1976, Ahmadinejad attended the Iran University of Science and Technology, where he received a bachelor's and a master's degree in civil engineering and in 1987, his doctorate in traffic and transportation planning. After receiving his degrees, Ahmadinejad became a lecturer in civil engineering and planning. In 1986, at the outbreak of the Iran-Iraq War, he became a member of the Islamic Revolutionary Guard Corps. He began his political career in 2003 when he was elected as the mayor of Tehran. In 2005,

Ahmadinejad was elected the sixth president of the Islamic Republic of Iran. In addition to being the president of a large oil-producing country, he is a significant figure in contemporary energy dynamics because of his promotion of Iran's nuclear development. Following a brief moratorium, in August 2005, after Ahmadinejad's election, Iran's nuclear facilities resumed uranium enrichment. This action spurred European nations and the United States to pressure the International Atomic Energy Agency (IAEA) to report Iran's nuclear facilities to the United Nation's Security Council, an action that the IAEA took in February 2006. But Iran did not halt nuclear development, and on April 11, 2006, Ahmadinejad announced that Iran had successfully enriched uranium that was suitable to use in a nuclear reactor for power generation. Since this announcement, tensions have existed between the United States, the European Union, and Iran. Ahmadinejad's role in energy dynamics is significant because it represents the dilemma that exists with nuclear energy. Iran has stated that its interest in nuclear technology is peaceful and it has every right to develop the power resources of its country. However, many developed nations are threatened by Iran's obtaining nuclear power, citing concerns of weapon development. This worry is likely to be an ongoing debate as more developing countries turn to nuclear sources to secure their energy needs.

John Browne, Lord Browne of Madingley (1948–present)

John Browne is the business leader of the world's largest oil company, BP. He was born in Hamburg, Germany, in 1948 and became introduced to the oil industry at a young age when his father went to work for Anglo-Persian Oil (later British Petroleum and BP). He received a bachelor's degree in physics from Cambridge University and a master's degree from Stanford University. He began his career at BP when he was still attending the university. Between the years 1969 and 1983, Browne oversaw production operations in Alaska, California, New York, the United Kingdom, and Canada. In 1986, he joined the Standard Oil Company of Ohio as vice president and chief financial officer. Standard Oil and British Petroleum merged in 1989, becoming BP.

Browne was appointed chief executive production officer in 1991, and in 1995, he accepted the position of group chief executive. Browne is a significant leader in the energy business because of his recognition of future energy trends. In 1997, Browne stated his intentions of transitioning BP into the "green" energy business. He acknowledged the problem of global warming as a pressing issue and called for energy companies to address growing energy demand with the development of renewable energy resources. In 1999, BP invested in renewable energy with its purchase of the solar energy company Solarex. It also has investments in wind power. Browne remains as the group CEO of BP and has announced that he will resign the post in December 2008.

Gro Harlem Brundtland (1939–present)

Gro Harlem Brundtland is an influential doctor and politician from Norway. She is noted in the area of energy dynamics for her promotion of the concept of sustainable development. Born in Oslo, Norway, in 1939, she developed a passion for political activism and medicine. She received her doctorate in medicine at the University of Oslo in 1963 and worked at the Norwegian Ministry of Health until 1974, when she was offered a position as minister of the environment. In 1981, she became the first woman to hold the office of prime minister in Norway, serving from 1986–1989 and again from 1990–1996. In addition to her service in public office, Brundtland made a significant contribution to global energy dynamics when she developed and chaired the UN World Commission of Environment and Development in 1983. Also known as the Brundtland Commission, this United Nations working group was charged with the task of examining the link between the environment and the global economy. The final report produced from this investigation in 1987 was titled *Our Common Future*. It introduced the concept of sustainable development as a new way of thinking about economic growth and development. The impact of this idea was important for the global environmental movement. It offered ideas for the global economy that would allow developing countries to promote economic and energy development in ways that were environmentally sustainable. Brundtland retired from her position as prime minister of Norway in 1996. In 1998, she became the director-general of the World Health Organization.

Lázaro Cárdenas (1895–1970)

Lázaro Cárdenas was one of the most influential political leaders in Mexican history. Born in Jiquilpan, Michoacán, in 1895, he was able to complete only six years of formal education before his father passed away, leaving him to care for his mother and siblings. He began his career in civil and military service in 1913 during the Mexican Revolution. In 1928, he became the governor of the Mexican state of Michoacán. He became the president of Mexico in 1934. Many of Cárdenas's administrative efforts involved constructing a modern democracy in Mexico that worked to the advantage of working-class people. His most significant contribution to global energy dynamics was the expropriation of Mexico's petroleum industry. After a series of failed negotiations with foreign oil companies, Cárdenas nationalized Mexico's oil industry and expropriated the property of seventeen foreign oil companies that operated in the country. He formed the nationalized oil company Pemex to assume operations of the oil industry in Mexico. This action was significant for world energy dynamics because it marked the first time that a country assumed control of its oil and natural gas industry from foreign entities. The Mexico appropriation served as an example worldwide that oil exporting countries had the ability to control their own oil resources. Cárdenas's presidential term ended in 1940. He served as the secretary of defense until 1945. He died of cancer in 1970 in Mexico City.

Andrew Carnegie (1835–1919)

Andrew Carnegie was an influential business leader in the industrial revolution. Born in Dunfermline, Scotland, in 1835, Carnegie was raised in poverty. His family immigrated to Pittsburgh, Pennsylvania, in 1848. Once in Pittsburgh, Carnegie began to build his fortune by working in a variety of trades, including the railroad and iron industries. He is most noted in energy history for his role in the steel industry, where he became a leader in steel refining and manufacturing. In 1865, he founded the Carnegie Steel Company, which soon became the world's largest steel company. The thriving steel industry was vital to the development and expansion of the railroad in the late 1800s, allowing for the rapid settlement of the western United States. It also transformed the eastern United States, creating a surging demand for coal. (Recall that steel was produced primarily from coke, a carbonized substance made from the pyrolysis of coal.) Coal mining increased in part to supply the burgeoning steel industry, and many eastern U.S. cities (notably Pittsburgh, where Carnegie had established his empire) became choked with coal-smoke pollution. In 1900, Carnegie sold his steel company to J. P. Morgan for $480 million. Throughout the remainder of his life, he used much of his fortune for philanthropic activities, creating the Carnegie Corporation, an organization devoted to promoting education in the United States, and establishing over 2,000 free public libraries in the United States and Europe. He died in Massachusetts in 1919.

Hugo Chávez (1954–present)

Hugo Chávez, the current president of Venezuela, is a notable, contemporary person in global energy affairs. He was born in 1954 in the town of Sabaneta, in the Venezuelan state of Barinas. He graduated with a degree in military arts and sciences in 1975 from the Venezuelan Academy of Military Sciences. Chávez began a military career that lasted for seventeen years, during which he established the Revolutionary Bolivarian Movement, which sought to overthrow the presidency of Carlos Andrés Pérez. After a failed coup d'état in 1992, Chávez was imprisoned for a short period and later pardoned. He was elected president of Venezuela in 1998. Upon arriving in office, he implemented programs to combat poverty and promote social development and oppose the rise of global market liberalism. During his presidential terms, Chávez has faced immense criticism and accusations of human rights violations. He is influential in the scope of global energy dynamics for his strong positions promoting the nationalization of Venezuela's oil industry. As a member of the Organization of Petroleum Exporting Countries (OPEC), Chávez has pushed for cartel-wide reductions in oil production as a mechanism for increasing global energy prices. He has also sought to renegotiate oil contracts with Exxon-Mobil operations in Venezuela. These actions have made him unpopular with many consumer nations, most notably the United States. His position in the global oil market is notable because it serves as an example of the link between foreign relations, national energy security, and conflict.

William Knox D'Arcy (1849–1917)

William Knox D'Arcy was the founder of the first foreign oil company in the Middle East. Educated at Westminster in London, he moved to Australia in 1866, where he founded the Mount Morgan Gold Mining Company. He returned to England with his family in 1889. In 1900, he began financing oil exploration expeditions to Persia (modern-day Iraq), where oil was struck by his ventures in 1908. He founded the Anglo-Persian Oil Company (APOC) in 1909 and began exporting oil resources from Persia. D'Arcy's role in oil development in the Middle East is notable for two reasons. First, his ventures were the first oil exploitation operations that occurred in the most oil-rich region in the world. Second, he founded the world's largest oil company. APOC later became British Petroleum (BP), an oil giant that continues to develop petroleum resources and operations in countries worldwide and whose record profits in 2006 made it the largest global oil company as measured by production. D'Arcy became a wealthy man through his investments in the oil industry. He died in 1917.

Thomas Edison (1847–1931)

Thomas Edison was born in Milan, Ohio. Although he became partially deaf in childhood, Edison began his career as a telegraph operator in the 1860s. Edison established himself as an inventor in 1877 when he patented the phonograph, a device designed to record and reproduce sound. In the late 1870s, Edison founded the first industrial research laboratory at Menlo Park in New Jersey to develop industrial technologies. It was at this facility that Edison and his research team invented the first commercially practical incandescent electric lightbulb by placing a carbonized filament in a vacuum and using electricity to produce light. This invention proved to be a major step in the widespread use of electric lighting. In 1878, Edison founded the Edison Electric Light Company, the first electric utility company. In 1882, he used incandescent lighting to operate at the Pearl Street Station, the first electric utility company. The company utilized direct current (DC) (as opposed to alternating current [AC], the current system) for electric power distribution. Although much of Edison's work was carried out by his research team at Menlo Park, Edison is an important figure in energy history because of his contribution to the cheap and efficient distribution of electric lighting for homes and businesses. This accomplishment paved the way for the numerous electric utility companies that would follow in his path. Prior to his death in 1931, Edison coordinated the first commuter electric train system in the United States, a line that stretched from Hoboken to Dover in New Jersey. He passed away in 1831 at the age of eighty-four in New Jersey.

Albert Einstein (1879–1955)

Albert Einstein is perhaps the most well-known physicist worldwide. His theories and ideas are notable in energy history for their contribution to the field of electromagnetism. Einstein was born in Württemberg, Germany, in 1879. He received a formal education in physics and mathematics at the Swiss Federal Polytechnic School, receiving a diploma in 1901 and his doctorate in 1905. In Switzerland, he published his special theory of relativity, which challenged traditional Newtonian physical laws. In 1914, he moved to Germany, where he became a professor at the University of Berlin. It was also during this year that he published his general theory of relativity, a document that contains his most renowned work. Einstein's theories formed the basis for the field of quantum mechanics. This discipline contributed to the fundamental laws of physics, allowing for a deeper understanding of the forces that operate at the subatomic level. Einstein's theories are important because they provide the foundation for the development of atomic theory and nuclear energy. His theories allowed for an understanding of nuclear dynamics. These ideas were conceptualized in the development of the atomic bomb and the harnessing of nuclear energy for society. In 1933, Einstein emigrated to the United States, where he accepted a position as professor of theoretical physics at Princeton University. He achieved a number of awards for his work, including the Nobel Prize for Physics in 1921, the Copley Medal from the Royal Society of

London in 1925, and the Franklin Medal from the Franklin Institute in 1935. He died in Princeton, New Jersey, in 1955.

Michael Faraday (1791–1867)

Michael Faraday was an English chemist and physicist who is known for his contributions to the study of electricity and magnetism. He was born in Newington, Surrey, England, in 1791. Although he had little formal education, he was appointed as an assistant to the English chemist Humphry Davy at the Royal Institute of London in 1812. He initially worked as a chemist, but soon became interested in the phenomenon of electromagnetism. In 1821, Faraday discovered the principle of electromagnetic rotation, the theoretical principle for the electric motor. His most notable discovery relating to the history of energy is the concept of electromagnetic induction. This principle was the foundation for the development of transformers (a device that converts electric current from high to low voltages) and generators (a machine that converts mechanical energy into electricity). These developments paved the way for the efficient and reliable distribution of electricity to society because they allowed electricity to be generated using a wide variety of energy resources. Faraday made two other important discoveries in the field of electromagnetism: the magneto-optical effect, where magnetic forces affect light; and the concept of diamagnetism, where substances align themselves with a magnetic field. Faraday concluded from these experiments that magnetism is a property of matter. Faraday continued to experiment in science until his death in 1867.

Henry Ford (1863–1947)

Henry Ford was born on July 30 on a Michigan farm. During his childhood and adolescence, he demonstrated an interest in engineering and in 1879 left his family farm to become an apprentice machinist in Detroit. In 1896, Ford built his first car, a gasoline-powered vehicle named the Quadricycle. Ford continued to work on his vehicle design and founded the Ford Motor Company in 1903. He is notable in energy history for his role in expanding the transportation industry. In 1913, Ford devised the assembly line, a manufacturing method that promoted the specialization of one task by laborers who completed their task on a product moving along a connected belt, ultimately resulting in a finished product. This labor innovation allowed Ford's Model T vehicle to be produced cheaply and efficiently. Ford's labor organization was incredibly effective; it revolutionized the transportation industry, giving American families the ability to purchase inexpensive vehicles. By 1918, half of all the vehicles purchased in America were Model Ts. Another notable point about Ford was his philosophy of labor. He strongly advocated for fair wages to be paid to his employees, a policy that allowed for the effective implementation of the assembly line. During his career, Ford also dabbled in the aviation industry, creating the Ford 4AT Tri-Motor, first flown in 1926 and also the first aircraft designed to transport passengers. He died in Dearborn, Michigan, from a cerebral hemorrhage in 1947. He was eighty-three years old.

James B. Francis (1815–1892)

James Francis made an important contribution to energy history when he designed the Francis water turbine. He was born in England in 1815 and came to the United States in 1833 when he was eighteen years old. He gained experience in the water infrastructure industry when he got a job with the Locks and Canal Company in Massachusetts. He became chief engineer of this company in 1837. During his career, Francis was interested in improving the design of hydraulic systems. He achieved success and made an important contribution to the way that humans harness energy when he developed the Francis turbine. This water turbine utilized the basic design of the turbine, a device that directed water in an outward flow, causing it to spin. The Francis turbine operated in a similar fashion, but it harnessed more energy from falling water by using water pressure changes to spin the turbine faster. In an ideal installation, the turbine was placed in an area where water under high pressure entered the turbine and was released into a low pressure environment; hence the pressure change helped extract useable energy from falling water. The design converted potential energy in water to mechanical energy with 90 percent efficiency. It also allowed for efficient operation in a range of water flow conditions. Because of these aspects, the Francis turbine is the most widely used design for water turbines in the world. It is a significant contribution to energy use, as hydropower accounts for approximately 20 percent of electricity generation worldwide. Francis is also notable as a founder of the American Society for Civil Engineers in 1880. He died on September 18, 1892.

Albert Arnold Gore (1948–present)

A political figure and environmental activist, Al Gore is a notable figure in energy dynamics for his campaign to raise awareness about global warming. He was born in Washington, DC, in 1948. The son of a U.S. senator, Gore spent his childhood between Washington, DC, and a farm in Tennessee. He obtained a bachelor of art's degree in government in 1969 and enlisted in the army during the Vietnam War. He was granted an honorable discharge in 1971. Gore began his political career when he became a U.S. representative from the state of Tennessee in 1976. He held that office until 1984, when he was elected to the U.S. Senate. In 1993, Gore became vice president under President William Clinton. In 2000, he campaigned for the presidency of the United States, and although he won the nation's popular vote, he failed to win enough votes in the Electoral College and conceded the election after a contentious challenge. Gore is an important figure in global energy dynamics for his efforts to raise awareness and combat global warming. Throughout his service in Congress, Gore worked to promote sustainable energy programs and enact environmental education programs. He supported the enactment of the Kyoto Protocol by the U.S. Senate and cosponsored hearings on global warming. After his defeat in the 2000 election, Gore has devoted his time to global warming outreach and education. His efforts are documented in the film *An Inconvenient Truth* (2006).

Otto Hahn (1879–1968)

Otto Hahn was one of the most influential nuclear chemists in energy history. He was born in Frankfurt, Germany, in 1879. He received his doctorate in organic chemistry from the University of Marburg in 1901. In 1904, he accepted a position at the University College of London, where he discovered radiothorium, a new radioactive substance. Hahn continued his work in nuclear chemistry at McGill University in Montreal, where he discovered radioactinium. In 1907, Hahn returned to Germany as a lecturer at the University of Berlin. In 1938, Hahn made the most important discovery of his career and his largest contribution to energy history. In collaboration with Fritz Strassmann, he found that the element barium was produced when uranium atoms were bombarded with neutrons. Hahn and Strassmann had discovered nuclear fission, the main chemical process that occurs in a nuclear reaction. Their discovery paved the way for the development of the atomic bomb and nuclear energy. Hahn received the Nobel Prize in 1944 for his discovery of nuclear fission. Hahn continued research in the development and separation of new elements through the process of nuclear fission. He died in 1968.

Marion King Hubbert (1903–1989)

Marion King Hubbert was born in San Saba, Texas, in 1903. He earned his bachelor of science, master of science, and doctoral degrees in geophysics from the University of Chicago. While earning his degrees, Hubbert was employed as a geologist for the American Petroleum Company, a job that would introduce him to the dynamics of the oil and natural gas industries. He worked as a researcher for Shell Oil Company from 1943 to 1964. He left the oil industry to work as a senior researcher for the U.S. Geological Survey until 1976. He is notable in the field of geophysics for his demonstration of the phenomenon of "plasticity," a process by which extreme heat and pressure can transform rock masses deep in the surface of the earth. His main contribution to energy dynamics came in 1956 when he presented a theory to the American Petroleum Institute. The Peak Oil Theory predicted that U.S. oil production would peak in the 1960s and then decline in subsequent years. The prediction proved true in the 1970s as production began to fall; and in 1975, the National Academy of Sciences gave further credence to the theory when the organization noted its previous estimates of oil and gas reserves had been overstated. The Peak Oil Theory is significant in energy history because it represents the first time that a renowned scientist warned of the depletion of an important energy resource. Hubbert received many awards for his scientific work, notably the Geological Society of America's Arthur L. Day Medal in 1954 and the National Academy of Science's (NAS) Penrose Medal in 1973. He became the president of the NAS in 1962. He died on October 11, 1989. He was eighty-six years old.

Kenneth Lay (1942–2006)

Kenneth Lay was born into a poor family in Tyrone, Missouri, in 1942. He attended the University of Missouri and received a doctorate in economics from the University of Houston in 1970. Lay became introduced to the oil industry when he went to work as an economist for the Exxon Corporation in 1965. In the early 1970s, Lay worked as an energy regulator, first for the Power Commission and then as the undersecretary for energy issues at the U.S. Department of the Interior. In 1974, he returned to the private energy industry. Throughout the late 1970s and 1980s, he held executive positions in several oil and natural gas companies. In 1985, he formed the Enron Corporation from a merger of Houston Natural Gas and the natural gas company Inter-North. In 1986, he assumed the position of chief executive officer of Enron. Enron's success was unprecedented. The company became a powerful electricity and natural gas utilities company, employing over 21,000 people and trading in the markets of over 800 energy service products. Lay is an important figure in energy dynamics not only for Enron's success but also because of the company's collapse. In 2001, Enron's stock declined dramatically when it was revealed that fraudulent accounting practices had inflated the company's wealth. The company went bankrupt, and thousands of employees lost their pension benefits and stock options. Lay and several of his associates were put on trial for the collapse of Enron. During the proceedings, Enron's abuses of California's deregulated energy market were revealed. In 2006, Lay was found guilty on six charges of financial crimes. He would not live to receive sentencing; he died of congestive heart failure on July 5, 2006. The story of Ken Lay and the Enron Corporation is a notable occurrence in energy dynamics because it describes the problems that may occur in a deregulated energy market. Lay's (and his associates') actions resulted in rolling blackouts in California, the loss of pensions and retirement funds for thousands of workers, and overall public distrust of energy providers.

Mohammad Mossadegh (1882–1967)

Mohammad Mossadegh was an important political figure in Iranian history. He received a Ph.D. from Neuchatel University in Switzerland. He returned to Iran in 1914 to become a public servant and elected official and aligned himself with the nationalist party, a political organization that opposed foreign-owned operations in Iran. He served as the prime minister of Iran from 1951 to 1953. He is notable in energy history for his attempts to nationalize the country's oil resources. Soon after Mossadegh became prime minister, the Iranian parliament, under his direction, voted to adopt the Oil Nationalization Act and seize control of assets held by the Anglo-Iranian Oil Company (AIOC), a British firm. The British government responded by creating a blockade against Iran's oil shipments, and the AIOC removed British personnel from the country. Without a market for its oil resources and lacking the expertise to operate oil refineries, Iran fell into an economic crisis. Mossadegh's government was removed from power in 1953 during a coup organized by the U.S. Central Intelligence Agency. This event is notable in energy history because it illustrates the significance of energy resources in the foreign policy decisions of oil-dependent nations. Iran's democratically elected government was removed from power because of its threat to the global oil industry. Mossadegh was arrested and tried for treason. He spent the rest of his life under house arrest and died in 1967. After his overthrow, a monarchy was reestablished in Iran, and foreign operation of Iran's oil industry resumed in 1954.

Jawaharlal Nehru (1889–1964)

Jawaharlal Nehru was one of the most influential leaders in Indian history. Born on November 14, 1889, Nehru was the son of a prominent member of the Indian National Congress. He was educated in England, where he received a law degree from Cambridge University. He returned to India in 1912 and after practicing law for several years, became an influential political leader, campaigning for India's freedom from British rule. Nehru worked closely with Mahatma Gandhi in his campaign for social reform and independence. Although he was jailed several times in his early political career for his radical political views, Nehru became the president of the Indian National Congress in 1929. On August 1, 1947, India was granted independence from Great Britain and Nehru became India's first prime minister. Nehru is an influential leader in India's energy history. Soon after India's independence, he adopted an industrialization planning and development policy. He promoted the importance of increasing India's electricity and coal production as a means of protecting India's independence. India's energy planning initiatives included the doubling of electricity and coal production, tripling iron ore production, commencing oil drilling operations in the Indian Ocean, and instituting a program of nuclear development. Energy intensities in India dramatically increased during this period of time as petrochemical and electronics industries were developed. As a result of these economic and industrial policies, India became an increasingly productive nation. Because of Nehru's industrialization policies, India is now considered one the most rapidly developing nations, a status that has large implications for global energy use. Nehru held his position as prime minister until he died on May 27, 1964.

J. Robert Oppenheimer (1904–1967)

J. Robert Oppenheimer was an influential nuclear energy scientist. He was born in New York City on April 22, 1904. He received a degree in chemistry from Harvard College in 1925 and received a Ph.D. from Göttigen University in Germany in 1927. After receiving his doctorate, Oppenheimer returned to the United States where he worked as a professor of theoretical physics at the University of California, Berkeley. During his time at UC–Berkeley in the 1930s, he developed the first theories that hypothesized the existence of black holes. Oppenheimer is notable in energy history because he led the team of scientists that built and detonated the first atomic bomb. Known as the

Manhattan Project, this top secret assignment was carried out at the Los Alamos weapons lab in New Mexico. The atomic bomb was first exploded on July 16, 1945. Three weeks later, two atomic bombs would be dropped on the Japanese cities of Hiroshima and Nagasaki, ending World War II and forever changing the atmosphere of global political and strategic relations. After the development of the atomic bomb, Oppenheimer became interested in promoting the peaceful use of nuclear energy. He was appointed the first chairman of the General Advisory Committee to the Atomic Energy Commission in 1946. During his time as chairman, he discouraged the development of the hydrogen bomb, a bomb based on principles of nuclear fusion and more powerful than the atomic bomb. Because of his opposition to the H-bomb, he made several enemies in the U.S. government, and his security clearance was eventually revoked. He died in Princeton, New Jersey, of throat cancer in 1967.

Medha Patkar (1954–present)

Medha Patkar is one of the most influential environmental and social leaders in India. She was born December 1, 1954, in Bombay, India. Her father's involvement in India's independence movement introduced her to activism. She earned an M.A. degree in social work from the Tata Institute of Social Sciences. In 1988, she formed the Narmada Bachao Andolan (Save Narmada Movement), a grassroots movement that sought to halt the construction of dams in the Narmada River Valley in India. The Narmada Valley Development Project, designed as part of India's economic development and planning, was to build an extensive network of hydroelectric dams. Although intended for the advancement of India, the dams would displace thousands of people, mostly residing in poor tribal and peasant communities. They would also inundate hundreds of acres of forested lands. The Narmada Bachao Andolan was successful in raising global awareness about the negative consequences of dam construction. In 2001, the Gujarat High Court ordered the Indian government to reconsider the project. Patkar has been granted many awards for her work, most notably Amnesty International's Human Rights Defenders Award. Her achievements are significant in global energy dynamics because they have raised awareness about the negative consequences of large dam construction.

Roger Revelle (1909–1991)

Roger Revelle was one of the first scientific pioneers to study the phenomenon of global warming. He was born in Seattle, Washington, in 1909 and raised in Pasadena, California. He received a B.S. in geology from Pomona College in 1929 and a Ph.D. in oceanography from the University of California, Berkeley, in 1936. He worked as a professor with the Scripps Institute of Oceanography in San Diego. After service in the navy during World War II, Revelle returned to Scripps to serve as its director from 1951 to 1964. During this time, he became interested in the dynamics of atmospheric CO_2 and ocean environments. Revelle is notable in energy history for his extensive research in the area of global warming. In 1957, Revelle partnered with Hans Suess to propose that human contributions of CO_2 to the atmosphere could lead to increased global warming. He prompted the scientific community to begin monitoring CO_2 in the oceans, in terrestrial ecosystems, and in the atmosphere. Revelle's own research in the area focused on the uptake of atmospheric CO_2 in ocean ecosystems. His studies concluded that the rate of CO_2 sequestration in ocean waters was much lower than initially estimated. These results, coupled with data demonstrating rapidly increasing atmospheric CO_2 concentrations, provided the foundation for contemporary studies of climate change. Revelle received many awards for his work, including the William Bowie Medal from the American Geophysical Union in 1968 and the National Medal of Science in 1991. He died on July 15, 1991. He was eighty-two years old.

John D. Rockefeller (1839–1937)

John D. Rockefeller was the pioneer of the modern U.S. petroleum company. Rockefeller became interested in the oil industry after the first American oil strike in Pennsylvania. He built an oil refinery there in 1863 and began oil development operations. His operations grew, and by 1870, he had formed the Standard Oil Company, unique for its vertically integrated organization. Not only did the company operate the extraction and production facilities, it also controlled the petroleum transportation network by offering incentives to railway and pipeline operators. This arrangement allowed the company to own and manage all aspects of oil production and distribution and led to monopoly control over the industry. In 1882, the company formed the Standard Oil Trust, an organization owned by nine trustees that further influenced pricing structures for oil markets. Because of its influence in the oil market, the trust came under attack by the antitrust movement and was eventually dissolved in 1911. Although this business structure was challenged and eventually dissolved because of its monopolistic nature, the trust initiated the development of similar arrangements in other industries. Rockefeller is an important person in energy history because his business ventures in the oil industry established a precedent for energy industries. He promoted the early development of petroleum industries and created a business model pursued by other energy companies. In 1897, John D. Rockefeller retired his position as president of the Standard Oil Company and became a full-time philanthropist, establishing charitable organizations for education and medical research.

Zhu Rongji (1928–present)

Zhu Rongji is an important contemporary figure in the development of China's energy resources. He was born in 1928 in Changsha, Hunan Province, China. He attended the Qinghua University and graduated with a degree in electrical engineering in 1951. From 1952 to 1969, Rongji worked for the state planning commission, where he was charged with helping to implement the Great Leap Forward reforms. These policies, instituted by President Mao Zedong, called for a rapid industrialization of China's economy and an increase in production from China's coal industry. Rongji criticized the Great Leap Forward, stating that its goals were unattainable. When Deng Xiaoping rose to power in 1978, Rongji was selected as an adviser for economic reforms. In 1979, Rongji became the chief of the Bureau of Fuel and Power Industry within the State Economic Commission. In 1989, he became the mayor of Shanghai, a post that he held until 1991, when he became the director of the State Council Production Office. In 1998, Rongji became the premier of the People's Republic of China. Rongji is an important figure in Chinese energy history because of his role in advancing the Chinese economy, including improving energy sectors and increasing energy production and industrialization. Rongji's experience in energy sectors and knowledge of economics have facilitated China's rapid industrialization. This rapid industrialization is significant because it has substantially increased Chinese energy consumption. Rongji's service as premier ended in 2003.

Franklin D. Roosevelt (1882–1945)

Franklin Delano Roosevelt (FDR) was one of the most influential presidents of the United States. He was born in Hyde Park, New York, in 1882. He obtained a B.A. degree in history from Harvard University and attended New York's Columbia University, where he studied law. He began his political career in 1910, holding various state and federal offices. In 1932, he became the 32nd president of the United States. He was reelected in 1936, 1940, and 1944, becoming the only U.S. president to serve for more than two terms. FDR is an important figure in energy history because many of the policies enacted by his administration facilitated the growth of energy industries in the United States. During the 1930s, FDR enacted a large number of federal regulations designed to pull the country out of the Great Depression. Many of these policies were designed to improve the U.S. economic situation, but they directly influenced energy growth and consumption. For example, the development of the Civilian Conservation Corps (CCC) employed thousands of young men in the construction of new roads and dams—projects that expanded the transportation and electricity industries in the nation. He also supported the development of the Tennessee Valley Authority (TVA), a publicly owned and operated utility company that brought hydropower to communities in the southeastern United States, and the Rural Electrification Act, which mandated the expansion of the electricity grid to rural areas of the country. In the latter half of his presidential career, World War II broke out. The United States rapidly increased production of energy and wartime resources and joined the war effort in 1941. FDR's policies ensured the supply of adequate energy resources to the Allied forces. FDR died of a cerebral hemorrhage during his fourth term in the presidential office. He was sixty-three years old.

Kenule Beeson Saro-Wiwa (1941–1995)

Kenule Beeson Saro-Wiwa was a Nigerian-born environmental activist and writer who initiated a campaign to raise awareness about the environmental degradation occurring from Shell Oil Company's activities in the Niger Delta. Born in Bori, he was a member of the Ogoni tribe that occupied the Niger Delta, an oil-rich region of Nigeria. After receiving degrees from the Government College in Umuahia, the University of Ibadan, and completing graduate studies at the University of Nigeria, he began a career in 1968 as a public servant for the Rivers State, a government entity created during Nigeria's civil war. In 1973, he was released from his position and became a writer, creating a satiric television comedy *(Basi and Company)* and several novels, most notably *Sozaboy: A Novel in Rotten English* in 1985 and *On a Darkling Plain* in 1987. He became an environmental activist in 1991 when he founded the Movement for the Survival of the Ogoni People (MOSOP) for the purpose of raising awareness about the environmental damage and injustices caused by the Shell Oil Company in Nigeria. He also spoke against the Nigerian government, accusing it of genocide against the Ogoni people. In 1993, Shell halted its oil development efforts on Ogoni tribal land. In 1994, he was awarded the Right Livelihood Award, an international honor, for his activism. In May 1994, Saro-Wiwa and eight other tribal members were accused of aiding in the deaths of four Ogoni chiefs at a political rally. They were found guilty in what many have determined to be an unfair trial and executed by hanging in 1995.

Joseph Stalin (1879–1953)

Joseph Stalin was an influential leader of the Soviet Union. Born to a poor family in Gori, Georgia, a province of the Russian Empire, he was originally named Iosif Vissarionovich Dzhugashvili. He renamed himself Stalin, meaning "man of steel," in 1913. In 1894, Stalin began his education at the Russian Orthodox Tiflis Theological Seminary. He was expelled from the seminary in 1899 for his involvement in the Georgian Social-Democratic organization, a political movement that promoted Marxist ideology. After his expulsion, Stalin was a fighter in an underground political movement until 1917, playing key roles in the Bolshevik Revolution, overthrow of the Russian Empire, and establishment of the ruling Communist Party. After the death of Russian leader Vladimir Lenin, Stalin quickly rose to power and by the late 1920s was the ruling dictator of the Soviet Union. He is important in Russian energy history because during his reign Russia experienced rapid, state-centralized industrialization under the implementation of a series of five-year plans. During this period, energy intensity rose dramatically as the production of coal doubled and iron ore more than tripled in the first five-year period. The rapid industrialization of Russia was spurred by the use of brutal, forced labor and wasteful practices of energy production. Because of these practices, Stalin's regime was highly criticized for its human rights violations and its decimation of the environment. Although Russia rose to be one of the most productive nations in the world, it serves as an example of poor energy planning. Following World War II, Stalin implemented a nuclear program that not only developed weapons but also promoted the use of nuclear energy for electricity generation. Stalin remained the dictator of Russia until 1953, when he died of a cerebral hemorrhage.

Maurice Strong (1929–present)

Maurice Strong is one of the most influential figures in the global environmental movement. He was born in Manitoba, Canada, in 1929. In 1948, Strong began a career as an oil and gas executive, eventually becoming the president of the Power Corporation of Canada. In 1966, Strong resigned from this position to become the head of Canada's International Development Office. In 1970, he became a leader in the global environmental movement when he assumed the position of secretary general of the United Nation's Conference on Human Environment. In 1973, Strong became the first executive director of the newly created United Nations Environment Programme, a UN affiliation whose mission is to promote global environmental stewardship through development assistance, education, and collaboration. Strong is an important figure in global energy dynamics because he bridged the gap between environment and energy development at the United Nations and facilitated the spread of global environmentalism. His connection to the energy industry allowed him to understand the importance of energy and environmental issues. From 1976 to 1984, Strong returned to the private sector, where he became the chief executive officer (CEO) of Petro-Canada until 1978, after which he served as chairman of various development corporations. In 1992, he became the CEO of North America's largest utility company, Ontario Hydro. That same year, Strong returned to the United Nations, where he served as the secretary general of the 1992 United Nations Conference on Environment and Development, also known as the Earth Summit. He serves as the president of the Council of the United Nations University for Peace.

Nikola Tesla (1856–1943)

Nikola Tesla is most notable in energy history for his development of alternating current electrical systems. He was born in the Croatian town of Smiljan in 1856. He received formal education at the Polytechnic School in Graz and later attended the University of Prague. He began a career in electrical engineering in Europe before moving to the United States in 1884. He worked with Thomas Edison for a time; however, the two inventors did not get along, and in 1885, Tesla began working with George Westinghouse on an electrical distribution system that utilized alternating current (AC). The use of AC in electrical networks is Tesla's most notable contribution to energy history. This technology was first used in 1893 by the Westinghouse Electric Company to light the World Colombian Exposition in Chicago. AC proved to be a more efficient and effective way than Edison's direct current (DC) system to transport electricity in a grid, and it became the basis for most modern electricity distribution systems. In addition to the AC system, Tesla contributed to the development of generators and turbine design. He also demonstrated fluorescent lighting. He continued to promote ideas of electricity generation and turbine design throughout the remainder of his life. He died in New York City in 1943.

James Watt (1736–1819)

James Watt developed one of the most important machines for society—the modern steam engine, which allowed humans to more efficiently capture the energy from coal combustion. Born in

Greenock, Scotland, in 1736, Watt gained experience in carpentry and shipbuilding. In 1755, he moved to London to become an apprentice to instrument maker John Morgan. In 1756, Watt returned to Scotland to work as an instrument maker at the University of Glasgow. It was at this position that Watt developed his greatest contribution to energy history. In 1765, Watt redesigned a bench-scale model of the Newcomen steam engine by adding a separate condenser to the design. This innovation allowed for steam to be cooled and recycled in the engine, making the device more efficient and powerful. After his discovery, Watt worked to develop a large-scale engine to be used for the purpose of pumping water from coal mines. In 1773, Watt partnered with business industrialist John Boulton. They produced the Boulton-Watt steam engine in 1774. The engine was first used in an industrial setting by the Bentley Mining Company in March 1776. After patenting his engine design, Watt continued to work on improving efficiency in steam engines, inventing a rotating piston and a double-acting engine (a device that made use of the power created during both the upward and downward stroke of the piston). In addition to water pumping, Watt's engines were used in transportation and milling applications. Watt died in 1819 at the age of eighty-three. In his honor, the international unit for power was named the watt.

Frank Whittle (1907–1996)

Frank Whittle was an inventor whose design of the modern jet engine had large impacts for energy dynamics of the twentieth century. He was born in Earlsdon, England, in 1907. He attended Leamington College until 1923, when he left to join the Royal Air Force (RAF). In 1928, he graduated from officer training at the College of Cranwell after completing a thesis on the principle of motorjet operation. In 1929, Whittle developed the idea of using a gas turbine to create a jet propulsion effect in an airplane engine, a process that would allow airplanes to travel at faster speeds and higher altitudes. He patented this design in 1932. In 1935, Whittle received financial support to build a prototype engine using his design. Initial models failed, but in 1941, the engine was tested on the trial plane, E.28 Pioneer, which flew a distance of 200 yards. After the initial flight testing of the jet engine, interest grew for the technology. It was a significant invention in

regard to air travel, revolutionizing the development of commuter and war planes. In 1948, Whittle retired from the RAF. In
1953, he became an engineering specialist for Shell Oil Company,
where he worked on developing a more efficient pump for oil
drilling operations. In 1976, he was knighted in Great Britain.
Soon after his knighthood, Whittle moved to the United States,
where he became a researcher and professor for the U.S. Naval
Academy. He died in Baltimore, Maryland, in August 1996.

6

Data and Documents

Introduction

Understanding energy use in society requires the compilation of a broad array of facts and statistics. These numbers describe how much energy is harnessed, what resources are extracted and consumed, the energy services that are delivered to various end users, and the waste that is created from the energy structure. Because energy systems are an extremely large and vital part of every society, understanding the numbers associated with them can be confusing. The approach of this chapter is to dissect various energy statistics, providing the reader with a clearer understanding of their importance in the broader picture of energy and society.

The purpose of this chapter is twofold. First, it is intended to be a quick reference for global and U.S. energy statistics. Second, it is to be used in conjunction with previous chapters in this book to provide a graphical and schematic overview of energy trends and flows in society. To meet these purposes, a broad overview of energy production and consumption is presented first. Next, statistics are broken down according to each energy resource used by society. Third, trends in global energy trade and carbon dioxide emissions from fossil fuel consumption are examined. Finally, an overview of energy use in the United States is provided. It is important to note that the statistics in this chapter represent a small fraction of the vast amount of data that is collected for energy analyses. The numbers reported here were chosen because they characterize the fundamental facts of energy use worldwide.

Energy Overview

Worldwide energy use is usually broken down into two statistics: primary energy production and primary energy consumption. These two figures offer a useful description of the total amount of energy harnessed by society, but sometimes the terminology used to generate these numbers can be confusing. Primary energy calculations generally represent the amount of energy contained in raw fuels (e.g., petroleum or coal), but do not include values for electricity generation and consumption, which is considered a secondary energy resource. However, to account for resources that do not have thermal energy equivalents (e.g., wind), information agencies often include electricity generated from hydropower and other renewable sources in their primary production and consumption figures. It is important to consider definitions when using energy statistics.

Table 6.1 provides a basic overview of primary energy production and consumption for each country worldwide. Countries are organized into seven geographic regions. These regional divisions are used throughout the rest of the chapter to illustrate global energy trends in the various statistics presented. Population data for each country are also provided. This table is intended to provide the reader with an overall picture of energy use for each nation and region worldwide.

Although the information in table 6.1 is useful for illustrating how much primary energy is produced and consumed in society at a given time, the values represent only a snapshot of energy use in 2004. These statistics can also be helpful for understanding how energy use has changed over time. Figure 6.1 illustrates how energy production and consumption have evolved in each world region over a period of twenty-four years (1980–2004). A couple important trends can be found from analyzing these graphs. First, there is a large gap in magnitude of energy use between certain regions. For example, Africa and Central and South America are two regions where more energy is produced than consumed, energy production and consumption are relatively low, and these values have remained fairly

TABLE 6.1

Global Total Energy Production, Consumption, and Population by Country and Region (2004)

Africa

Country	Population (millions)	Production (10^{15} Btu)	Consumption (10^{15} Btu)	Country	Population (millions)	Production (10^{15} Btu)	Consumption (10^{15} Btu)
Algeria	32.13	7.144	1.239	Madagascar	17.50	0.006	0.036
Angola	11.52	2.284	0.141	Malawi	12.41	0.013	0.024
Benin	7.44	0.00002	0.030	Mali	11.13	0.002	0.011
Botswana	1.64	0.024	0.053	Mauritania	3.00	0.0004	0.051
Burkina Faso	13.09	0.001	0.018	Mauritius	1.22	0.002	0.055
Burundi	7.52	0.001	0.008	Morocco	32.21	0.024	0.444
Cameroon	16.64	0.181	0.086	Mozambique	19.11	0.119	0.139
Cape Verde	0.42	0	0.002	Namibia	2.01	0.014	0.055
Central African Republic	4.17	0.001	0.006	Niger	11.81	0.005	0.017
Chad	9.38	0.390	0.003	Nigeria	125.74	5.901	1.012
Comoros	0.65	0.00002	0.002	Reunion	0.77	0.006	0.044
Congo (Brazzaville)	3.50	0.502	0.017	Rwanda	8.24	0.001	0.013
Congo (Kinshasa)	58.92	0.116	0.087	Saint Helena	0.01	0	0.0002
Cote d'Ivoire (Ivory Coast)	16.94	0.142	0.110	Sao Tome and Principe	0.18	0.0001	0.001
Djibouti	0.47	0	0.026	Senegal	11.43	0.002	0.067
Egypt	76.12	2.829	2.523	Seychelles	0.08	0	0.012
Equatorial Guinea	0.52	0.747	0.006	Sierra Leone	5.73	0	0.014
Eritrea	4.55	0	0.011	Somalia	8.30	0	0.010
Ethiopia	71.34	0.023	0.084	South Africa	44.45	6.065	5.119
Gabon	1.36	0.528	0.040	Sudan	39.15	0.648	0.148

continued

TABLE 6.1, continued

Global Total Energy Production, Consumption, and Population by Country and Region (2004)

Country	Population (millions)	Production (10^15 Btu)	Consumption (10^15 Btu)
Africa			
Gambia	1.55	.0000	0.004
Ghana	21.48	0.063	0.142
Guinea	9.23	0.005	0.023
Guinea-Bissau	1.39	.0000	0.005
Kenya	32.98	0.056	0.175
Lesotho	2.04	0.003	0.005
Liberia	2.81	.0000	0.007
Libya	5.63	3.609	0.749
Swaziland	1.14	0.011	0.021
Tanzania	36.07	0.024	0.072
Togo	5.26	0.003	0.034
Tunisia	9.97	0.270	0.333
Uganda	26.39	0.019	0.041
Western Sahara	0.27	.0000	0.004
Zambia	11.03	0.104	0.122
Zimbabwe	12.08	0.156	0.204
Total	**872.11**	**32.043**	**13.706**
Asia and Oceania			
Afghanistan	28.51	0.006	0.016
American Samoa	0.06	.0000	0.008
Australia	19.91	10.555	5.266
Bangladesh	141.34	0.477	0.658
Bhutan	2.19	0.022	0.020
Brunei	0.37	0.851	0.100
Burma	46.52	0.479	0.202
Malaysia	23.52	4.096	2.519
Maldives	0.34	.0000	0.015
Mongolia	2.75	0.070	0.093
Nauru	0.01	.0000	0.002
Nepal	27.07	0.024	0.063
New Caledonia	0.21	0.003	0.028
New Zealand	3.99	0.681	0.884

continued

TABLE 6.1, continued

Global Total Energy Production, Consumption, and Population by Country and Region (2004)

Asia and Oceania

Country	Population (millions)	Production (10^{15} Btu)	Consumption (10^{15} Btu)
Cambodia	13.40	0.0004	0.008
China	1,298.85	55.948	59.573
Cook Islands	0.02	.0000	0.001
East Timor	1.02	0.101	NA
Fiji	0.88	0.007	0.028
French Polynesia	0.27	0.001	0.013
Guam	0.17	.0000	0.034
Hong Kong	6.86	.0000	1.091
India	1,065.07	11.055	15.417
Indonesia	238.45	8.842	4.686
Japan	127.33	4.034	22.624
Kiribati	0.10	.0000	0.0004
Korea, North	22.70	0.852	0.891
Korea, South	48.43	1.355	8.985
Laos	6.07	0.046	0.050
Macau	0.45	.0000	0.032
Niue	—	.0000	0.00004
Pakistan	159.20	1.395	1.986
Papua New Guinea	5.42	0.108	0.075
Philippines	86.24	0.498	1.310
Samoa	0.18	0.0004	0.002
Singapore	4.35	.0000	1.936
Solomon Islands	0.52	.0000	0.003
Sri Lanka	19.91	0.029	0.197
Taiwan	22.75	0.471	4.399
Thailand	63.73	1.596	3.423
Tonga	0.11	.0000	0.002
U.S. Pacific Islands	0.26	0.0003	0.004
Vanuatu	0.20	.0000	0.001
Vietnam	82.66	1.574	0.948
Wake Island	—	.0000	0.019
Total	**3,572.39**	**105.177**	**137.613**

continued

TABLE 6.1, continued

Global Total Energy Production, Consumption, and Population by Country and Region (2004)

Central and South America

Country	Population (millions)	Production (10^{15} Btu)	Consumption (10^{15} Btu)	Country	Population (millions)	Production (10^{15} Btu)	Consumption (10^{15} Btu)
Antarctica	—	.0000	0.003	Guyana	0.76	0.0001	0.024
Antigua and Barbuda	0.07	.0000	0.008	Haiti	7.94	0.003	0.027
Argentina	39.14	3.756	2.788	Honduras	7.01	0.018	0.101
Aruba	0.07	.0000	0.014	Jamaica	2.71	0.002	0.158
Bahamas	0.30	.0000	0.057	Martinique	0.43	.0000	0.032
Barbados	0.28	0.003	0.024	Montserrat	0.01	.0000	0.001
Belize	0.27	0.001	0.014	Netherlands Antilles	0.22	.0000	0.153
Bolivia	8.72	0.509	0.198	Nicaragua	5.36	0.011	0.069
Brazil	184.10	7.210	9.078	Panama	3.09	0.038	0.209
Cayman Islands	0.04	.0000	0.005	Paraguay	6.19	0.519	0.420
Chile	15.82	0.306	1.181	Peru	27.54	0.418	0.577
Colombia	42.31	3.203	1.193	Puerto Rico	3.89	0.001	0.550
Costa Rica	3.96	0.096	0.186	Saint Kitts and Nevis	0.04	.0000	0.002
Cuba	11.31	0.178	0.469	Saint Lucia	0.16	.0000	0.006
Dominica	0.07	0.0003	0.002	Saint Vincent/Grenadines	0.12	0.0003	0.003
Dominican Republic	8.92	0.017	0.301	Suriname	0.44	0.037	0.038
Ecuador	13.21	1.242	0.383	Trinidad and Tobago	1.08	1.377	0.593
El Salvador	6.59	0.034	0.123	Turks and Caicos Islands	0.02	.0000	0.0002
Falkland Islands	—	.0000	0.0005	Uruguay	3.40	0.081	0.173
French Guiana	0.19	.0000	0.015	Venezuela	25.02	8.031	2.884

continued

TABLE 6.1, continued
Global Total Energy Production, Consumption, and Population by Country and Region (2004)

Country	Population (millions)	Production (10^{15} Btu)	Consumption (10^{15} Btu)	Country	Population (millions)	Production (10^{15} Btu)	Consumption (10^{15} Btu)
				Central and South America			
Grenada	0.09	.0000	0.004	Virgin Islands, U.S.	0.11	.0000	0.244
Guadeloupe	0.44	.0000	0.028	Virgin Islands, British	0.02	.0000	0.001
Guatemala	11.73	0.076	0.180				
Total	**443.22**	**27.169**	**22.517**				
				Eurasia			
Armenia	2.99	0.046	0.180	Lithuania	3.61	0.174	0.356
Azerbaijan	7.87	0.889	0.655	Moldova	4.45	0.003	0.127
Belarus	10.31	0.085	0.967	Russia	143.97	51.689	30.062
Estonia	1.34	0.117	0.223	Tajikistan	7.01	0.164	0.272
Georgia	4.69	0.068	0.143	Turkmenistan	4.86	2.617	0.808
Kazakhstan	15.14	4.932	2.331	Ukraine	47.31	3.244	6.486
Kyrgyzstan	5.08	0.141	0.170	Uzbekistan	26.41	2.516	2.227
Latvia	2.31	0.028	0.172				
Total	**287.36**	**66.714**	**45.179**				
				Europe			
Albania	3.54	0.071	0.114	Italy	58.09	1.303	8.265
Austria	8.17	0.521	1.456	Luxembourg	0.46	0.002	0.200
Belgium	10.35	0.501	2.784	Macedonia	2.04	0.058	0.112
Bosnia and Herzegovina	4.35	0.160	0.215	Malta	0.40	.0000	0.041

continued

TABLE 6.1, continued

Global Total Energy Production, Consumption, and Population by Country and Region (2004)

Country	Population (millions)	Production (10^{15} Btu)	Consumption (10^{15} Btu)	Country	Population (millions)	Production (10^{15} Btu)	Consumption (10^{15} Btu)
				Europe			
Bulgaria	7.52	0.389	0.846	Netherlands	16.32	2.940	4.103
Croatia	4.50	0.161	0.387	Norway	4.57	10.784	1.941
Czech Republic	10.25	1.137	1.770	Poland	38.58	3.041	3.667
Denmark	5.41	1.281	0.864	Portugal	10.52	0.123	1.111
Faroe Islands	0.05	0.001	0.011	Romania	22.36	1.146	1.644
Finland	5.21	0.471	1.346	Serbia and Montenegro	10.83	0.520	0.772
France	60.46	5.185	11.250	Slovakia (Slovak Republic)	5.42	0.286	0.797
Germany	82.42	5.358	14.693	Slovenia	2.01	0.151	0.330
Gibraltar	0.03	.0000	0.054	Spain	40.28	1.479	6.402
Greece	10.65	0.426	1.446	Sweden	8.99	1.449	2.317
Hungary	10.03	0.393	1.065	Switzerland	7.48	0.616	1.287
Iceland	0.29	0.100	0.148	Turkey	68.89	1.000	3.533
Ireland	3.97	0.044	0.637	United Kingdom	60.27	9.511	10.038
Total	**584.72**	**50.610**	**85.647**				
				Middle East			
Bahrain	0.68	0.457	0.414	Lebanon	3.78	0.009	0.238
Cyprus	0.78	.0000	0.114	Oman	2.90	2.258	0.374
Iran	67.50	12.050	6.449	Qatar	0.84	3.450	0.706

continued

TABLE 6.1, continued

Global Total Energy Production, Consumption, and Population by Country and Region (2004)

Country	Population (millions)	Production (10^15 Btu)	Consumption (10^15 Btu)	Country	Population (millions)	Production (10^15 Btu)	Consumption (10^15 Btu)
				Middle East			
Iraq	25.37	4.376	1.207	Saudi Arabia	25.80	24.159	6.100
Israel	6.20	0.029	0.873	Syria	18.02	1.258	0.824
Jordan	5.61	0.012	0.280	United Arab Emirates	2.52	7.424	2.336
Kuwait	2.26	5.706	1.061	Yemen	20.02	0.888	0.164
Total	**182.28**	**62.078**	**21.139**				
				North America			
Bermuda	0.06	.0000	0.008	Mexico	104.96	10.305	6.609
Canada	32.51	18.617	13.600	Saint Piere and Miquelon	0.01	.0000	0.001
Greenland	0.06	.0000	0.008	United States	293.03	70.388	100.414
Total	**430.62**	**99.310**	**120.641**				
World Total	**6,372.72**	**443.100**	**446.442**				

Source: U.S. Energy Information Administration, *International Energy Annual 2004*, International Data. http://www.eia.doe.gov/emeu/international/contents.html. Accessed September 25, 2006.

stagnant since 1980. In comparison, Europe, North America, and Asia and Oceania consume more energy than they produce and use considerably more energy overall.

A second important trend is the growth in energy production and consumption in the Asia and Oceania region, as evidenced by the steep slopes of their curves. This increase in energy use is indicative of the rapid industrialization that is occurring in countries like India and China (see chapter 2 for discussion). In comparison, other industrialized regions that have higher energy use values, like Europe and North America, have remained relatively static in their growth.

Two additional trends worth pointing out are seen in Eurasia and the Middle East. In Eurasia, there is an obvious dip in both energy production and consumption beginning around 1990 and persisting for about a decade. This decrease in energy use represents the fall of the Soviet Union (see chapter 2 for discussion). The Middle East, notably, is the only region where energy production is considerably greater than consumption. This trend supports the fact that the Middle East contains the largest reserves of the world's most traded energy commodity, petroleum. The Middle East chart shows a significant dip in production around 1982. This decline in production was the result of the Iran-Iraq War that began in 1980 and lasted until 1988.

Figure 6.1 demonstrates the utility of overall energy production and consumption data. The previous paragraphs pointed out how these statistics can be indicators for major global and regional events. However, as useful as these facts are, they only scratch the surface of what energy data can reveal. The numbers used in the previous table and figure lump together information gathered for all energy resources used by society; however, earlier chapters demonstrated that not all energy resources are created equal. The energy produced and consumed from individual resources must be considered to paint a complete energy picture. The next section breaks down total energy figures into their respective resource origins.

Energy Resources

In preceding chapters, it was noted that over 85 percent of the energy consumed worldwide is from fossil fuels. But that fact does not tell what specific resources are used in different regions.

FIGURE 6.1
Primary Energy Production and Consumption by Region (1980–2004)

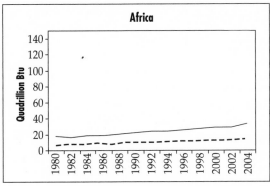

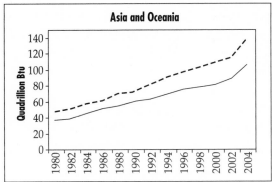

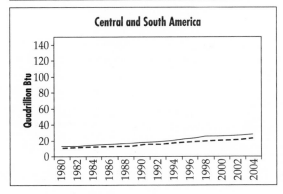

continued

FIGURE 6.1, continued
Primary Energy Production and Consumption by Region (1980–2004)

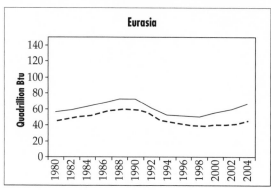

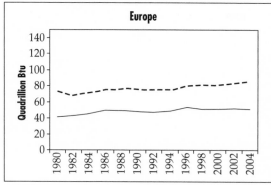

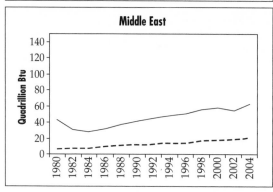

continued

FIGURE 6.1, continued
Primary Energy Production and Consumption by Region (1980–2004)

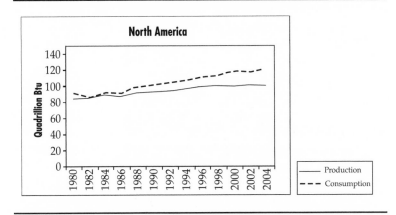

Source: Energy Information Administration (EIA), "International Data Tables, Total World Energy Production and Consumption (1980–2004)," *International Energy Annual 2004.*

Figure 6.2 provides a more comprehensive breakdown of energy use. While fossil fuels comprise a large slice of the pie in all regions, certain fossil fuels are more important in some regions than others. For example, in Asia and Oceania, coal accounted for 49 percent of the total energy consumption in 2004. In the Middle East, petroleum takes the largest part of the pie, comprising 54 percent of energy consumption. And in Eurasia, natural gas is the dominant fossil fuel, making up 53 percent of the total energy consumed. These regional differences can be useful for understanding what resources are indigenous to particular areas in the world. From this figure, one is able to gain a better understanding of global energy use, but further questions remain. How much of each resource does each region produce and consume? What countries consume and produce the most oil, coal, or natural gas? And what are the particular pathways that an energy resource must travel in order to benefit society?

The next section examines the facts associated with particular energy sources and how each resource is exploited by society. Energy cycle charts that highlight important flows of energy, where the resource comes from, and the various steps involved in the delivery of energy services and products from each resource complement the statistics.

FIGURE 6.2
Regional Primary Energy Consumption by Fuel Type (2004)

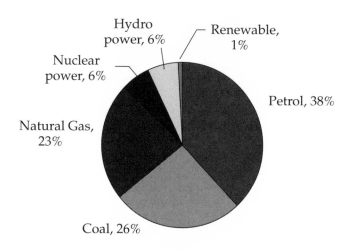

World

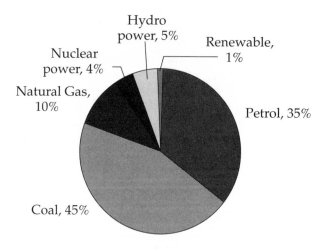

Asia and Oceania

continued

FIGURE 6.2, continued
Regional Primary Energy Consumption by Fuel Type (2004)

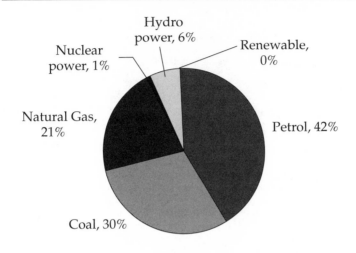

Africa

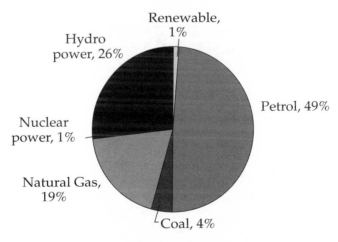

Central and South America

continued

FIGURE 6.2, continued
Regional Primary Energy Consumption by Fuel Type (2004)

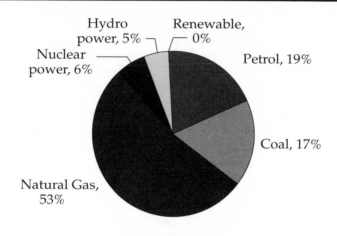

Eurasia

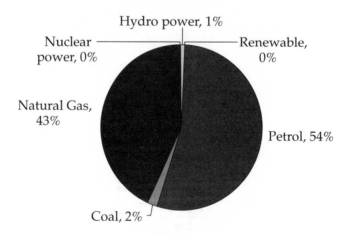

Middle East

continued

FIGURE 6.2, continued
Regional Primary Energy Consumption by Fuel Type (2004)

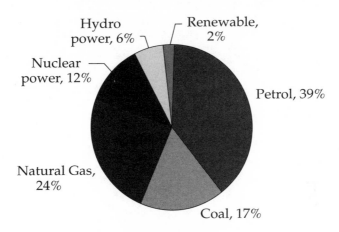

Hydro power, 6%

Renewable, 2%

Nuclear power, 12%

Petrol, 39%

Natural Gas, 24%

Coal, 17%

Europe

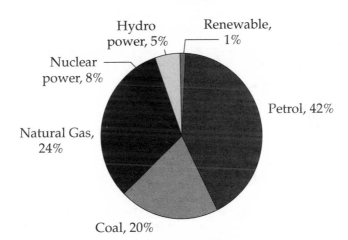

Hydro power, 5%

Renewable, 1%

Nuclear power, 8%

Petrol, 42%

Natural Gas, 24%

Coal, 20%

North America

Source: Energy Information Administration (EIA), "International Data," *International Energy Annual 2004,* http://www.eia.doe.gov/emeu/international/contents.html. Accessed September 25, 2006.

Fossil Fuels

Petroleum, natural gas, and coal make up 87 percent of the energy consumed worldwide. The flow path of each of the three primary fossil fuels, from extraction and production to consumption, is illustrated in figures 6.3–6.5. These energy flow diagrams illustrate the larger picture that is involved in the production and consumption of fossil fuels. From these diagrams, a few important but often overlooked aspects in each process can be identified. For example, market reports state the amount of crude oil produced, or the price for a barrel of crude oil. The emphasis on this aspect of the petroleum cycle overlooks the process of refining. Figure 6.3 shows that refining is the crucial step required for the delivery of marketable petroleum products for consumers, most notably gasoline and jet fuel.

Figure 6.4 demonstrates that although natural gas resources are often found with petroleum resources, the gas is not always used. Instead, natural gas extracted with oil is sometimes reinjected into the reservoir, vented, or flared. Natural gas flaring, often disregarded by market reports, can be very polluting and dangerous. For this reason, it is considered illegal in many producing countries.

Finally, figure 6.5 demonstrates the path of coal from extraction to consumption. An important consideration of coal mining is the amount of refuse waste that is produced. This waste is often separated after the resource is mined and stored in large coal waste impoundments. Since the refuse can be up to 50 percent of the coal mined, it can pose a problem if it is not stored properly. Another important point to make regarding the production of coal is its importance in the steel industry. Coking coal remains one of the most widely utilized fuels in steel production, making it a vital resource in the process of industrialization.

Tables 6.2 and 6.3 provide statistics for fossil fuel production and consumption for each world region in 2004. These data are presented in physical units for each resource. These tables provide a snapshot of global energy use, but they do not consider how each resource has been used over time. Figure 6.6 fills this gap by graphing the consumption of each fossil fuel over the period of twenty-four years (1980–2004). From this figure, one is able to identify resource trends that may have an impact on regional energy dynamics. For example, in five out of the seven regions, petroleum is the dominant resource consumed. This trend

FIGURE 6.3
Petroleum's Cycle

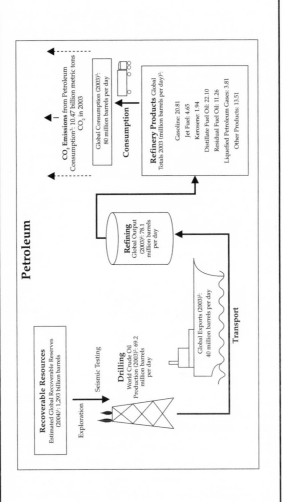

Sources: 1. PenWell Corporation, 2006, *Oil & Gas Journal,* September 227 (9), obtained from the Energy Information Administration (EIA), "International Petroleum Reserves Data."
2. EIA. *International Energy Annual,* table posted online June 19, 2006, http://www.eia.doe.gov/emeu/international/oilproduction.html. 3. EIA, *International Energy Annual* 2004,
http://www.eia.doe.gov/emeu/international/contents.html. Accessed November 2, 2006.

FIGURE 6.4
Natural Gas' Cycle

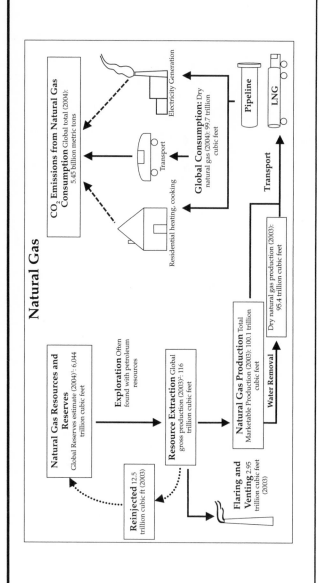

Sources: 1. PennWell Corporation, 2004, *Oil & Gas Journal* December 102 (47). Obtained from Energy Information Administration (EIA). http://www.eia.doe.gov/iea 2. EIA. "World Natural Gas Production, 2003." In *International Energy Annual 2004.* http://www.eia.doe.gov/iea. Accessed November 2, 2006.

FIGURE 6.5
Coal's Cycle

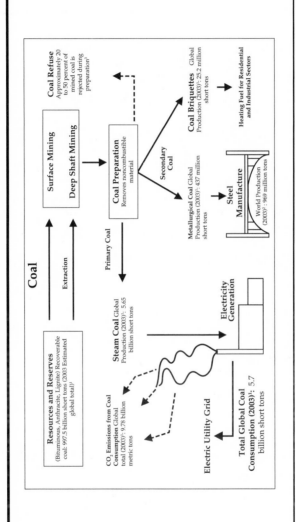

Sources: 1. Energy Information Administration (EIA), *International Energy Annual 2004*, http://www.eia.doe.gov/emeu/international/contents.html. 2. World Coal Institute, *Coal and Steel Facts 2006*, http://www.worldcoal.org/pages/content/index.asp?PageID=189. 3. National Academy of Sciences, 2002, *Coal Waste Impoundments: Risks, Responses and Alternatives* (Washington, DC: National Academy Press), 23. Accessed November 2, 2006.

TABLE 6.2
Fossil Fuel Production by Region (2004)

Region	Crude oil (million barrels per day)	Percent global total	Natural gas (trillion cubic feet)	Percent global total	Coal (thousand short tons per week)	Percent global Total
Africa	8.79	12.2	5.28	5.4	274	4.5
Asia and Oceania	7.43	10.3	12.10	12.3	3,232	53.2
Central and South America	6.12	8.6	4.54	4.6	75	1.2
Eurasia	10.53	14.6	28.16	28.6	493	8.1
Europe	5.72	7.9	11.89	12.1	806	13.3
Middle East	22.37	31.0	9.95	10.1	1.1	0.02
North America	11.20	15.5	26.70	27.1	1,197	19.7
Total	**72.22**		**98.62**		**6,079**	

Source: Energy Information Administration (EIA), "Petroleum, Coal and Natural Gas Data Tables," *International Energy Annual 2004,* http://www.eia.doe.gov/emeu/international/contents.html. Accessed October 10, 2006.

TABLE 6.3
Fossil Fuel Consumption by Region (2004)

Region	Crude oil (million barrels per day)	Percent global total	Natural gas (trillion cubic feet)	Percent global total	Coal (thousand short tons per week)	Percent global total
Africa	2.79	3.4	2.62	2.6	205.83	3.4
Asia and Oceania	23.34	28.3	13.47	13.5	3,190.25	52.3
Central and South America	5.38	6.5	4.08	4.1	38.21	0.6
Eurasia	4.11	5.0	23.39	23.5	429.40	7.0
Europe	16.31	19.7	19.90	20.0	1,036.30	17.0
Middle East	5.66	6.9	8.61	8.6	16.27	0.3
North America	25.00	30.3	27.60	27.7	1,182.53	19.4
Total	**82.59**		**99.67**		**6,098.78**	

Source: Energy Information Administration (EIA), "Petroleum, Coal and Natural Gas Data Tables," *International Energy Annual 2004,* http://www.eia.doe.gov/emeu/international/contents.html. Accessed October 10, 2006.

FIGURE 6.6
Regional Petroleum, Natural Gas, and Coal Consumption (1980–2004)

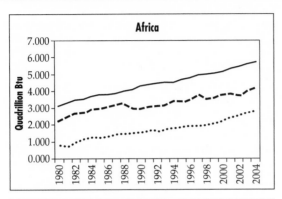

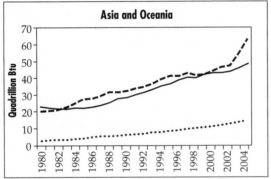

continued

FIGURE 6.6, continued
Regional Petroleum, Natural Gas, and Coal Consumption (1980–2004)

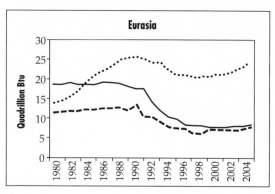

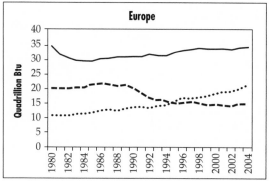

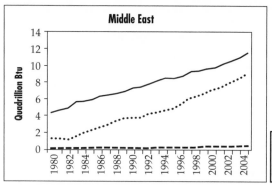

continued

FIGURE 6.6, continued
Regional Petroleum, Natural Gas, and Coal Consumption (1980–2004)

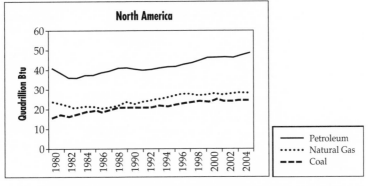

Source: Energy Information Administration (EIA), "International Data Tables," *International Energy Annual 2004,* http://www.eia.doe.gov/emeu/international/contents.html. Accessed October 10, 2006.

is important when considering future energy needs. As noted in previous chapters, petroleum production is projected to decline as resources become depleted. Resource depletion will make it more difficult for regional economies that are structured on petroleum consumption.

As meaningful as regional trends are, they do not provide detail on how much energy individual countries utilize. This distinction is important because often one or two countries in a particular region are responsible for a large share of resource production or consumption. For example, the United States comprises a large share of the energy consumed in North America. South Africa is also notable as it consumes five times the energy of all other African nations combined. Tables 6.4–6.6 show what resources countries produce and consume the largest amounts of.

Electricity

Electricity is the most important secondary energy resource that is used by society. Generally, values for electricity are divided into three categories: electric capacity, electricity generation, and

TABLE 6.4
Top Ten Petroleum-Producing and -Consuming Countries

Rank	Total Petroleum* (thousand barrels per day)			
	Country	Production	Country	Consumption
1	Saudi Arabia	10,492.6	United States	20,731.2
2	Russia	9,273.7	China	6,400.0
3	United States	8,700.2	Japan	5,353.2
4	Iran	4,101.7	Russia	2,770.0
5	Mexico	3,847.6	Germany	2,649.9
6	China	3,635.4	India	2,450.0
7	Norway	3,196.6	Canada	2,294.0
8	Canada	3,135.2	Korea, South	2,148.7
9	Venezuela	2,854.8	Brazil	2,140.0
10	United Arab Emirates	2,760.1	France	1,977.2

* Total petroleum = crude oil, natural gas plant liquids and other liquids, and refinery processing gain
Source: Energy Information Administration (EIA), "International Data Tables," *International Energy Annual 2004*, http://www.eia.doe.gov/emeu/international/contents.html. Accessed October 25, 2006.

TABLE 6.5
Top Ten Natural Gas-Producing and -Consuming Countries

Rank	Natural Gas (trillion cubic feet)			
	Country	Production	Country	Consumption
1	Russia	22.386	United States	22.430
2	United States	18.757	Russia	16.022
3	Canada	6.483	Germany	3.576
4	United Kingdom	3.389	United Kingdom	3.477
5	Netherlands	3.036	Canada	3.385
6	Iran	2.963	Ukraine	3.051
7	Norway	2.948	Iran	3.021
8	Algeria	2.830	Japan	2.950
9	Indonesia	2.663	Italy	2.847
10	Saudi Arabia	2.319	Saudi Arabia	2.319

Source: Energy Information Administration (EIA), "International Data Tables," *International Energy Annual 2004*, http://www.eia.doe.gov/emeu/international/contents.html. Accessed October 25, 2006.

TABLE 6.6
Top Ten Coal-Producing and -Consuming Countries

Rank	Coal (million short tons)			
	Country	Production	Country	Consumption
1	China	2,156.38	China	2,062.39
2	United States	1,112.10	United States	1,107.25
3	India	443.72	India	478.16
4	Australia	390.96	Germany	279.95
5	Russia	308.88	Russia	257.52
6	South Africa	267.67	Japan	203.72
7	Germany	232.67	South Africa	195.14
8	Poland	177.70	Poland	153.10
9	Indonesia	142.31	Australia	150.09
10	Kazakhstan	95.70	Korea, South	90.56

Source: Energy Information Administration (EIA), "International Data Tables," *International Energy Annual 2004,* http://www.eia.doe.gov/emeu/international/contents.html. Accessed October 25, 2006.

consumption. Capacity refers to the maximum amount of power that can be supplied to the electricity grid from a generating unit. This value is generally expressed in the units of megawatts. Recall from chapter 1 that a watt is an energy unit that describes the rate at which electricity can be generated; one watt is equivalent to a generation rate of one joule per second. A megawatt equals a rate of a million joules per second. Hence an electric capacity of 500 megawatts means that a particular generating unit has the ability to produce 500 million joules of energy per second in the form of electricity. A 500-megawatt plant can produce enough energy to serve about 250,000 households. A watt already has time factored into it, so 250,000 homes would be served continuously (or at least as long as the operational life of the power plant). It is important to note that generators do not always produce at capacity.

Electricity generation is often divided into values of gross and net generation. Gross generation calculates the actual amount of electricity that was produced from a generating unit. Net electricity generation subtracts the amount of electricity that is used in the operation of the generating facility from the gross value. Both values are generally reported in units of kilowatt-hours. Electricity consumption is the amount of electricity that reaches the end user. It is typically slightly less than the value of generation to account for transmission losses. Table 6.7 provides an overview of these three electricity statistics by world region.

TABLE 6.7
Electricity Capacity, Generation, and Consumption by Region (2004)

Region	Installed Capacity (thousand megawatts)	Net Electricity Generation (billion kilowatt-hours)	Net Electricity Generation (billion kilowatt-hours)
Africa	103.7	505.4	469.6
Asia and Oceania	1,074.6	5,103.0	4,748.4
Central and South America	212.2	881.4	819.6
Eurasia	342.4	1,307.3	1,194.8
Europe	781.4	3,440.0	3,217.5
Middle East	112.0	566.6	526.8
North America	1,110.1	4,795.4	4,464.6
Total	**3,736.3**	**16,599.1**	**15,441.3**

Source: Energy Information Administration (EIA), "International Data, Total Electric Generation, 1980–2004," *International Energy Annual 2004,* http://www.eia.doe.gov/emeu/international/electricitygeneration.html. Accessed October 25, 2006.

FIGURE 6.7
World Electric Capacity by Fuel Type (2004)

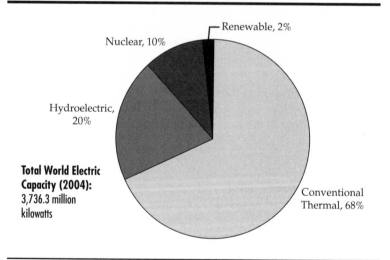

Total World Electric Capacity (2004): 3,736.3 million kilowatts

Source: Energy Information Administration (EIA), "International Data, International Electricity Generation Tables," *International Energy Annual 2004,* http://www.eia.doe.gov/emeu/international/electricitygeneration.html. Accessed October 25, 2006.

Electricity is generated from a variety of sources. Figure 6.7 presents a pie chart that breaks down world electric capacity by fuel type. Conventional thermal resources refer to generating units that utilize fossil fuels. They account for the largest amount of installed electric capacity, followed by hydropower, nuclear, and renewable resources.

Nuclear

Nuclear power is an interesting facet of energy dynamics in terms of both global relations and its growth potential as a vital energy resource in the future. The nuclear fuel cycle is illustrated in figure 6.8. It is important to point out that the process of nuclear fuel production and the construction of reactors can be cost prohibitive. The main limiting step in this process for many countries is the ability to enrich uranium. Producing nuclear fuel is often tainted with the concern that countries will also attempt the production of nuclear weapons. Once a country has established a viable nuclear fuel cycle, the operation of a generating facility pays off. It is cheap to produce electricity because nuclear fuel contains more potential energy per unit mass than any of the fossil fuels. Another positive aspect is that nuclear power plants do not release greenhouse gases. However, the nuclear cycle incorporates other waste issues, most notably the large mill tailing piles resulting from uranium fuel processing and the radioactive waste produced from spent fuel rods.

In 2004, nuclear energy accounted for 6 percent of the total energy used worldwide (see figure 6.2). According to the International Atomic Energy Agency (IAEA), there are thirty-one countries that produce nuclear power, operating a total of 442 commercial reactors. These countries are listed in table 6.8, which details the number of reactors currently operating in each country, the electric capacity of the generating units of these reactors, and the net electric power generated in 2005.

Figure 6.9 breaks down nuclear electricity generation by region. This pie chart shows that of the seven world regions, only four of them produce a sizeable amount of nuclear energy. South Africa is the only country in the region of Africa that produces electricity, and only two countries, Brazil and Argentina, have nuclear capabilities in Central and South America.

FIGURE 6.8
Nuclear Fuel's Cycle

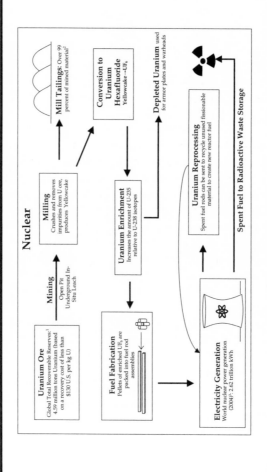

Sources: 1. H. Holger Rogner, 2004, "Uranium," in *2004 Survey of Energy Resources* (Oxford, England: Elsevier), 157. 2. E. Willard Miller and Ruby M. Miller, 1993, *Energy and American Society* (Santa Barbara, CA: ABC-CLIO), 43. 3. International Atomic Energy Agency (IAEA), 2006, "Operational and Long-Term Shutdown Reactors by Country," *Power Reactor Information System* (PRIS), http://www.iaea.org/programmes/a2/. Accessed November 2, 2006.

TABLE 6.8
Nuclear Reactors, Generation, and Capacity by Country (2005)

Country	Number of Nuclear Reactors[1]	Nuclear Electric Capacity (MW(e))[1]	Net Nuclear Power Generation (billion kWh) (2004)[2]	Country	Number of Nuclear Reactors[1]	Nuclear Electric Capacity (MW(e))[1]	Net Nuclear Power Generation (billion kWh) (2004)[2]
United States	103	98,145	788.53	Switzerland	5	3,220	25.61
France	59	63,363	425.83	Bulgaria	4	2,722	15.60
Japan	55	47,593	271.58	Finland	4	2,676	21.55
Russia	31	21,743	137.47	Slovakia	6	2,442	16.18
Germany	17	20,339	158.97	Brazil	2	1,901	11.60
Korea, South	20	16,810	124.18	South Africa	2	1,800	14.28
Ukraine	15	13,107	82.69	Hungary	4	1,755	11.32
Canada	18	12,584	85.87	Mexico	2	1,360	8.73
United Kingdom	23	11,852	73.68	Lithuania	1	1,185	14.35
Sweden	10	8,916	73.43	Argentina	2	935	7.31
China	10	7,587	47.95	Slovenia	1	656	5.21
Spain	8	7,450	60.43	Romania	1	655	5.27
Belgium	7	5,801	45.80	Netherlands	1	450	3.63
Taiwan	6	4,884	37.94	Pakistan	2	425	1.93
India	16	3,483	15.04	Armenia	1	376	2.21
Czech Republic	6	3,373	25.01				
Total	**442**	**369,588**	**2,619.18**				

Sources: 1. International Atomic Energy Agency, 2006, "Operational and Long-Term Shutdown Reactors by Country," Power Reactor Information System (PRIS), http://www.iaea.org/programmes/a2/.
2. Energy Information Administration (EIA), "International Electricity Data, Net Nuclear Power Generation," *International Energy Annual 2004*, http://www.eia.doe.gov/emeu/international/electricitygeneration.html. Accessed November 2, 2006.

FIGURE 6.9
Nuclear-Electricity Generation by Region (2004)

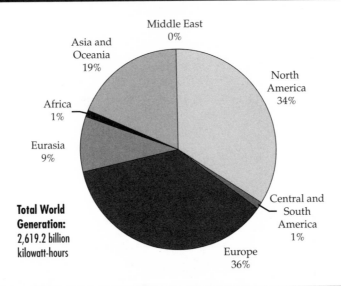

Source: Energy Information Administration (EIA), 2004, "International Electricity Data, Net Nuclear Power Generation," *International Energy Annual 2004*, http://www.eia.doe.gov/emeu/international/electricitygeneration.html. Accessed November 2, 2006.

Renewable Energy

Hydroelectric, solar, wind, biomass, and geothermal resources are the five categories of renewable energy resources depicted in figure 6.10. With the exception of geothermal energy, solar radiation provides most of the energy that drives the renewable cycles. The facts in this section focus largely on the potential of renewable sources to generate electricity from hydroelectric, solar, wind, and geothermal resources. Statistics for biomass are difficult to determine because often these resources are not commercially traded. It is important to point out that renewable sources can also provide a variety of direct energy uses (i.e., not electricity generation).

FIGURE 6.10
Renewable Energy's Cycle

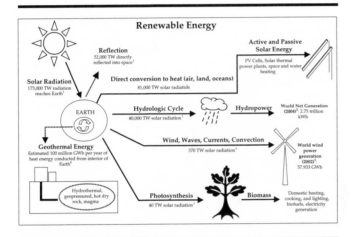

Sources: 1. Gary Alexander, "Overview: The Context of Renewable Energy Technologies," in Godfrey Boyle, ed. *Renewable Energy: Power for a Sustainable Future* (Oxford: Oxford University Press), p. 28.
2. Energy Information Administration, *International Energy Annual 2004,* International Electricity Generation Tables. Table posted July 7, 2006. http://www.eia.doe.gov/emeu/international/electricitygeneration.html. Accessed November 4, 2006.
3. Søren Varming, "Wind Energy," in *2004 Survey of Energy Resources: 20th Edition.* World Energy Council (Amersterdam: Elsevier), p. 69.
4. Lucien Bronicki and Michael Lax. "Geothermal Energy," in *2004 Survey of Energy Resources 20th Edition.* World Energy Council (Amersterdam: Elsevier), p. 346–348.

Hydropower accounts for 6 percent of the energy consumed worldwide (figure 6.2). It is of greatest importance in developing regions of the world. For example, in 2004, hydropower accounted for 26 percent of the energy consumed in Central and South America. Table 6.9 details the hydroelectricity generated worldwide in 2004 by region. Despite opposition to large dams and diversion projects, the development of hydropower resources is likely to grow in the future.

There are three primary end uses for solar energy: photovoltaic (PV) cells for the generation of electricity, solar thermal systems used for hot water and space heating, and large-scale solar thermal power plants. Because there are a variety of end uses, and because solar energy is often not commercially traded, it is difficult to monitor how much total energy is being harnessed from solar radiation on a global scale. Additionally, many solar technologies are still being developed. Consequently, they are not as readily available as fossil fueled technologies. Because of these factors, statistics on solar energy use are limited. The technology is growing, however, and many companies are exploring the market potential of PV cells. Table 6.10 details the top ten PV manufacturers in the world. The World Energy Council estimated that in 2002 there was approximately 1,500 MW of PV capacity installed globally (Silvi 2004, 298). Since PV cells are used in both grid-connected and nongrid-connected systems, it is impossible to determine the total amount of power generated from this technology.

TABLE 6.9
World Hydroelectricity Capacity, Generation, and Consumption by Region (2004)

Region	Hydroelectric Capacity (GW)	Hydroelectric generation (billion kWh)	Hydroelectric consumption (billion kWh)
Africa	21.068	87.43	87.43
Asia and Oceania	198.181	664.00	664.00
Central and South America	122.224	577.08	577.08
Eurasia	66.920	233.09	233.09
Europe	167.205	543.56	543.56
Middle East	6.499	14.11	14.11
North America	157.488	627.61	627.61
World Total	**739.585**	**2,746.88**	**2,746.88**

Source: Energy Information Administration (EIA), 2006, "International Electricity Generation Tables," *International Energy Annual 2004,* http://www.eia.doe.gov/emeu/international/electricitygeneration.html. Accessed November 6, 2006.

TABLE 6.10
Top Ten Manufacturers of Photovoltaic Solar Cells

Rank	PV Manufacturer	Country	Capacity (MW) (2002)
1	Sharp	Japan	123.1
2	BP Solar	United Kingdom	73.8
3	Kyocera	Japan	60.0
4	Shell	United Kingdom	57.5
5	Sanyo Electric	Japan	35.0
6	AstroPower	United States	29.7
7	RWE Solar	Germany	29.5
8	Isofoton	Spain	27.4
9	Mitsubishi Electric	Japan	24.0
10	Photowatt	France	17.0

Source: Cesare Silvi, 2004, "Solar Energy." In *2004 Survey of Energy Resources,* World Energy Council 298. Amsterdam: Elsevier. 20th ed.

TABLE 6.11
Wind-Electric Capacity and Generation by Region (2002)

Region	Installed Capacity (MWe)	Electricity Generation (GWh)
Africa	144	430
Asia	2,627	5,379
Central and South America	132	441
Oceania	144	496
Middle East	20	51
North America	4927	12460
Europe (including Eurasia)	23,404	38,676
World Total	**31,398**	**57,933**

Source: Søren Varming, "Wind Energy." In *2004 Survey of Energy Resources,* World Energy Council 369. Amsterdam: Elsevier. 20th ed.

Wind energy is also becoming an important renewable resource for regions looking to harness renewable sources of energy. In 2002, there was a total of 31,400 MW of installed electric capacity from wind energy (Varming 2004, 364). Table 6.11 provides data on the regional capacity and generation of wind energy worldwide in 2002. Table 6.12 examines the top ten wind energy producing countries for that same year.

TABLE 6.12
Top Ten Wind-Power-Generating Countries (2002) Ranked by Capacity

Rank	Country	Installed Capacity (MWe)	Annual Output (GWh)
1	Germany	12,001	16,800
2	Spain	4,825	9,792
3	United States	4,685	12,000
4	Denmark	2,889	4,877
5	India	1,702	3,700
6	Italy	788	1,600
7	Netherlands	693	1,200
8	United Kingdom	552	1,450
9	China	468	1,000
10	Japan	415	598

Source: Søren Varming, "Wind Energy." In *2004 Survey of Energy Resources,* World Energy Council 369. Amsterdam: Elsevier. 20th ed.

TABLE 6.13
Geothermal Electric and Direct-Use Capacity by Region (2002)

Region	Installed Electrical Capacity (MWe)	Installed Direct Use Capacity (MWt)
Africa	57	121
Asia	3,332	4,284
Central and South America	374	46
Oceania	448	318
Middle East	0	216
North America	2,855	5,908
Europe (including Eurasia)	1,154	6,107
World Total	**8,220**	**17,000**

Source: Lucien Bronicki and Michael Lax, "Geothermal Energy," in *2004 Survey of Energy Resources,* World Energy Council 346–348. Amsterdam: Elsevier. 20th ed.

Geothermal energy is the final category of resources discussed in this chapter. There are a total of twenty-one countries that utilize geothermal resources for both electricity generation and direct uses, such as heating. Figure 6.10 depicts four types of geothermal resources: hydrothermal, geopressurized, hot dry rock (also known as enhanced geothermal systems), and magma. Of these, hydrothermal resources are the most utilized. Table 6.13 provides an overview of both the installed electric and direct use capacities by region in 2002.

Energy Trade

The data described in the previous sections provides the types and amounts of energy produced and consumed. Yet a large part of energy use in society is dependent upon the global trade of these resources, the companies that exist in global energy markets, and the consequences of how these resources are distributed. This section briefly examines these topics.

Fossil fuels are the most traded commodity worldwide, with petroleum at the top of the list. There is not one country in the world that can survive without importing or exporting a portion of its resources. Tables 6.14 and 6.15 describe the top ten importers and exporters of oil, coal, and natural gas in 2004. It is interesting to note that the United States is the number one importer of petroleum and natural gas. More detailed information on the U.S. petroleum supply is presented later in the chapter.

Energy commodities are traded and delivered by hundreds of thousands of companies worldwide. Most of these companies operate on a global scale; that is, they extract and produce energy in several countries rather than just their country of origin. Table 6.16 lists the top twenty-five global energy companies of 2005

TABLE 6.14
Top Ten Importers and Exporters of Crude Oil (2004)

	Top Importers		Top Exporters
Country	Net Imports (million barrels/day)	Country	Net Exports (million barrels/day)
United States	12.1	Saudi Arabia	8.73
Japan	5.3	Russia	6.67
China	2.9	Norway	2.91
Germany	2.4	Iran	2.55
Korea, South	2.2	Venezuela	2.36
France	1.9	United Arab Emirates	2.33
Italy	1.7	Kuwait	2.20
Spain	1.6	Nigeria	2.19
India	1.5	Mexico	1.80
Taiwan	1.0	Algeria	1.68

Sources: Importers: Energy Information Administration (EIA), "Non-OPEC Fact Sheet," http://www.eia.doe.gov/emeu/cabs/topworldtables3_4.html; exporters: EIA, "Non-OPEC Fact Sheet," http://www.eia.doe.gov/emeu/cabs/topworldtables1_2.html. Accessed October 10, 2006.

TABLE 6.15
Top Ten Importers and Exporters of Coal and Natural Gas (2004)

Rank	Natural Gas (billion cubic ft)				Coal (trillion Btu)			
	Country	Imports	Country	Exports	Country	Imports	Country	Exports
1	United States	4,259	Russia	7,656	Japan	4,415.7	Australia	5,324.2
2	Germany	3,182	Canada	3,673	Korea, South	1,867.6	China	2,939.0
3	Japan	2,868	Norway	2,663	Taiwan	1,499.0	Indonesia	2,169.8
4	Italy	2,398	Algeria	2,150	Germany	1,035.7	South Africa	1,988.4
5	Ukraine	2,373	Netherlands	1,891	United Kingdom	831.0	Russia	1,358.3
6	France	1,581	Turkmenistan	1,483	United States	694.5	Colombia	1,234.4
7	Russia	1,293	Indonesia	1,353	Canada	629.3	United States	1,135.1
8	Korea, South	1,022	Malaysia	1,040	India	606.9	Canada	811.5
9	Spain	952	United States	854	Italy	589.5	Poland	678.4
10	Turkey	767	Qatar	850	Russia	575.0	Kazakhstan	456.7

Source: Energy Information Administration (EIA), 2006, "International Data Tables," *International Energy Annual 2004*, http://www.eia.doe.gov/emeu/international/contents.html. Accessed October 10, 2006.

ranked by their profits. This list was compiled by Platt's, an energy information company that provides yearly rankings of the top 250 energy companies. Although Platt's considers companies in nine industrial categories, most of those represented in table 6.16 are from the integrated oil and gas (IOG) sector. In fact, of the thirty-two IOG companies found in the top 250, seventeen of them are in the top twenty-five. This point highlights the importance of oil and natural gas commodities in the global market. Other classifications not on the table are the coal and consumable fuels, gas utility, independent power producers, and storage and transfer industrial sectors.

TABLE 6.16
Top Twenty-five Global Energy Companies (2005)

Rank	Company	Country	Industry	Date Established	Profits (million)
1	Exxon-Mobil Corp.	United States	Oil and Gas	1870	$36,130
2	Royal Dutch Shell	Netherlands	Oil and Gas		$25,618
3	BP plc	United Kingdom	Oil and Gas	1889	$22,157
4	Total	France	Oil and Gas	1924	$14,940
5	ConocoPhillips	United States	Oil and Gas	1917	$13,640
6	PetroChina Co. Ltd.	China	Oil and Gas	1988	$16,521
7	Chevron Corp.	United States	Oil and Gas	1879	$14,099
8	Petroleo Brasileiro SA	Brazil	Oil and Gas	1953	$10,582
9	ENI SpA	Italy	Oil and Gas	1953	$10,698
10	Statoil ASA	Norway	Oil and Gas	1972	$4,736
11	Valero Energy Corp.	United States	Refining and Marketing	1955	$3,590
12	Marathon Oil Corp.	United States	Oil and Gas	1887	$3,051
13	Occidental Petroleum Corp.	United States	Oil and Gas	1920	$5,272
14	China Petroleum	China	Oil and Gas	2000	$4,248
15	LUKoil Co.	Russia	Oil and Gas	1993	$4,248
16	Repsol YPF SA	Spain	Oil and Gas	1986	$3,798
17	Electricite de France	Spain	Electric Utility	1987	$3,947
18	Centrica plc	United Kingdom	Diversified Utility		$1,720
19	Gazprom OAO	Russia	Oil and Gas	1993	$7,298
20	Oil and Natural Gas Corp. Ltd.	India	Exploration and Production	1956	$3,209
21	Norsk Hydro AS	Norway	Exploration and Production	1905	$2,422
22	E.On AG	Germany	Electric Utility	1929	$5,331
23	ENEL SpA	Italy	Electric Utility	1962	$3,193
24	Imperial Oil Ltd.	Canada	Oil and Gas	1880	$2,237
25	Electrabel SA	Belgium	Electric Utility	1905	$2,323

Source: Platt's, 2005, "Top 250 Global Energy Company Rankings," http://www.platts.com/top250/index.xml. Data used with permission from Platt's.

A final, important thing to point out about energy trade is distribution inequality: a large gap exists between rich and poor nations. Recall the discrepancy in energy use shown in figure 6.1, which demonstrates that Africa and Central and South America use far less energy than North America and Europe. This point can also be made in relation to world economics.

Table 6.17 shows energy consumption per capita and gross domestic product (GDP) of the world's regions and some selected countries. Energy consumption per capita is a widely used indicator of energy dynamics. It describes how much energy is used by each person per year in a given country or region. The indicator accounts for an individual's share of energy resources ex-

TABLE 6.17
Energy and Economic Indicators by Region and Selected Country

Region and Selected Countries	Energy Consumption per capita (million Btu) (2004)	GDP (billion U.S. dollars, purchasing power parity) 2005
Africa	**15.7**	**2,365**
Nigeria	8.1	174.1
South Africa	115.2	533.2
Asia and Oceania	**38.5**	**22,644**
China	45.9	8,859
India	14.5	3,611
Japan	177.7	4,018
Central and South America	**50.8**	**3,515**
Brazil	49.3	1,556
Venezuela	115.3	153.7
Eurasia	**157.2**	**2,408**
Russia	208.8	1,589
Europe	**146.5**	**13,668**
France	186.1	1,816
Germany	178.3	2,504
United Kingdom	166.5	1,830
Middle East	**116.0**	**1,546.9**
Saudi Arabia	236.5	338
North America	**280.2**	**14,546.9**
Canada	418.4	1,114
United States	342.7	12,360
World Total	**70.1**	**60,710**

Sources: Energy consumption per capita: Energy Information Administration (EIA), "International Data," *International Energy Annual 2004,* http://www.eia.doe.gov/emeu/international/contents.html; GDP data: *CIA World Factbook,* 2006, http://www.cia.gov/cia/publications/factbook/index.html. Accessed October 13, 2006.

FIGURE 6.11
Global CO$_2$ Emissions from Fossil Fuels (1800–2003)

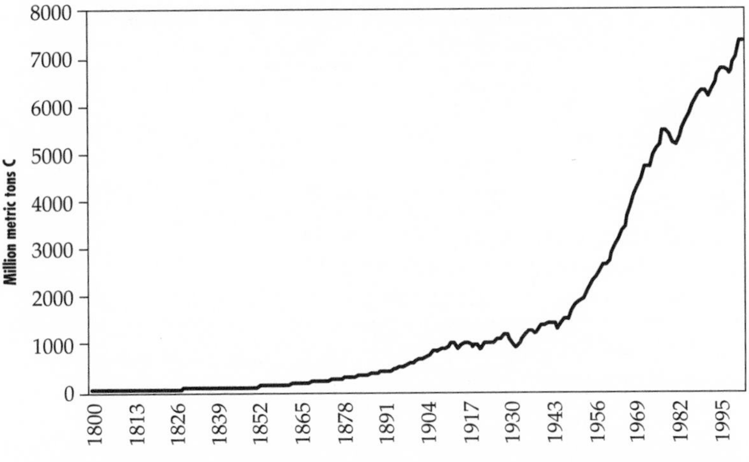

Source: G. Marland, T. A. Boden, and R. J. Andres, 2006, "Global, Regional, and National CO$_2$ Emissions," in *Trends: A Compendium of Data on Global Change, Carbon Dioxide Information Analysis Center* (Oak Ridge, TN: Oak Ridge National Laboratory, U.S. Department of Energy), http://cdiac.ornl.gov/trends/emis/tre_glob.htm. Accessed October 1, 2006.

pended. The main point of this table is to demonstrate that richer countries and regions (those with a higher GDP) typically consume more energy per capita than those in poorer regions.

Environment

Earlier chapters discuss problems associated with energy use. They note that the most pressing of these issues is global warming caused from carbon dioxide emissions that are released during the combustion of fossil fuels. This section provides information and data about these emissions.

It has been estimated that approximately 305 billion tons of carbon has been released into the atmosphere from fossil fuel consumption and cement production since 1791 (Marland et al. 2006). Figure 6.11 illustrates this increase since 1800. As this figure

TABLE 6.18
CO_2 Emissions from Fossil Fuels by Region (2004)

Region	CO_2 Emissions from Fossil Fuels (million metric tons)	CO_2 Emissions per capita (metric tons)	CO_2 Emissions from Fossil Fuels (million metric tons, carbon equivalent)
Africa	986.55	1.13	269.06
Asia and Oceania	9,604.81	2.69	2,619.49
Central and South America	1,041.45	2.35	284.03
Eurasia	2,550.75	8.88	695.66
Europe	4,653.43	7.96	1,269.12
Middle East	1,319.70	7.24	359.92
North America	6,886.88	15.99	1,878.24
World Total	**27,043.57**	**4.24**	**7,375.20**

Source: Energy Information Administration (EIA), 2006, "International Data," *International Energy Annual 2004*, http://www.eia.doe.gov/emeu/international/contents.html. Accessed October 20, 2006.

TABLE 6.19
Top Ten CO_2 Emitters (2004)

Rank	Country	CO_2 emissions (million metric tons)	Percent of World Total
1	United States	5,912.21	21.9
2	China	4,707.28	17.4
3	Russia	1,684.84	6.2
4	Japan	1,262.10	4.7
5	India	1,112.84	4.1
6	Germany	862.23	3.2
7	Canada	587.98	2.2
8	United Kingdom	579.68	2.1
9	Korea, South	496.76	1.8
10	Italy	484.98	1.8

Source: Energy Information Administration (EIA), 2006, "International Carbon Dioxide Emissions and Carbon Intensity, Table" *International Energy Annual 2004*, http://www.eia.doe.gov/emeu/international/carbondioxide.html. Accessed October 20, 2006.

FIGURE 6.12
Global CO$_2$ Emissions by Fuel Type (2004)

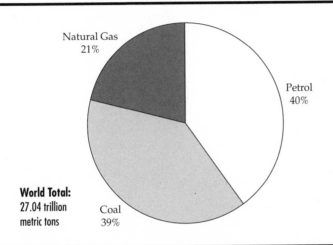

Natural Gas
21%

Petrol
40%

World Total:
27.04 trillion
metric tons

Coal
39%

Source: Energy Information Administration (EIA), 2006, "International Carbon Emissions and Intensity" tables, *International Energy Annual 2004*, http://www.eia.doe.gov/emeu/international/carbondioxide.html. Accessed October 20, 2006.

shows, a dramatic increase in CO$_2$ emissions has occurred in the latter half of the twentieth century. It is this trend that concerns scientists and policymakers studying the issue of global climate change.

Carbon dioxide emission data may be reported in various ways. The two most common approaches estimate total CO$_2$ emissions and the calculated carbon equivalent that is released from other CO$_2$ and other greenhouse gases. Values for total CO$_2$ emissions are relatively straightforward; they provide a value for the total amount of CO$_2$ released when a fossil fuel is burned. Carbon equivalents are a little more complicated. These measures are a standardized way of calculating the amount of total carbon that is released from any greenhouse gas relative to its warming potential, or its ability to trap solar radiation. Table 6.18 provides an overview of the amount of carbon dioxide that was released from the combustion of fossil fuels in 2004 by region. Table 6.19 ranks the top ten CO$_2$ emitters. Finally, figure 6.12 depicts the percent of CO$_2$ emissions by fuel type.

U.S. Data

The remaining tables and graphs in this chapter provide an overview of energy use in the United States. This section is included for two main reasons. First, the data in this section complement the information presented in chapter 3. Second, the United States is the world's largest consumer of energy, making it the most important player in global energy dynamics.

This brief summary of U.S energy statistics first provides total energy consumption for each state (table 6.20), and then presents data for the production of individual resources. Because different energy industries are distributed into different geographic regions, summaries of regional energy facts are difficult to compile. State divisions are included for different resources.

Where possible, energy data are described using U.S. Census Bureau divisions (table 6.21). Such is the case for the overall energy overview (figure 6.13). Information for natural gas production is also presented according to Census Bureau divisions, but the Gulf of Mexico offshore region is distinguished as its own category because it makes up 21 percent of U.S. natural gas production (figure 6.17).

Crude oil production is generally organized into PAD districts, so named for the districts drawn by the Petroleum Administration for Defense (PAD) in 1950 for purposes of supply management after World War II (table 6.22). Overall petroleum production from 1900 to 2005 is summarized in figure 6.14. An important note to make here is that U.S. petroleum production peaked in the late 1960s. Figure 6.15 depicts the most recent production numbers for the United States divided by PAD district. There are two additional figures in this section that summarize important petroleum data. Figure 6.20 shows U.S. petroleum import and export trends since 1960, it illustrates the dramatic increase in imports while exports remained relatively unchanged. Figure 6.21 complements the previous figure by illustrating the top ten foreign sources of petroleum for the United States.

Coal production is also categorized according to its own geographic regions (table 6.23). Figure 6.16 shows the amount of coal produced in the United States broken down by coal producing region. Here it is important to point out the prevalence of surface mining techniques utilized in the western region. In these

TABLE 6.20
U.S. Energy Overview by State and Region (2003)

State	Population	Consumption (trillion Btu)	State	Population	Consumption (trillion Btu)
Northeast					
Connecticut	3,485,881	888.7	New York	19,228,031	4,220.6
Maine	1,308,245	478.5	Pennsylvania	12,364,930	3,972.7
Massachusetts	6,417,565	1,588.8	Rhode Island	1,075,729	227.7
New Hampshire	1,287,594	327.5	Vermont	619,092	155.8
New Jersey	8,640,028	2,578.3			
Total	**54,427,095**	**14,438.6**			
Midwest					
Illinois	12,649,940	3,918.3	Missouri	5,718,717	1,841.8
Indiana	6,196,269	2,912.8	Nebraska	1,738,013	646.1
Iowa	2,941,362	1,175.8	North Dakota	633,051	395.0
Kansas	2,724,224	1,117.9	Ohio	11,431,748	3,986.2
Michigan	10,078,146	3,158.2	South Dakota	764,599	263.9
Minnesota	5,061,662	1,795.8	Wisconsin	5,471,792	1,832.5
Total	**65,409,523**	**23,044.3**			
South					
Alabama	4,501,862	2,013.5	Mississippi	2,880,793	1,183.8
Arkansas	2,726,166	1,132.8	North Carolina	8,422,375	2,643.7
Delaware	817,827	312.9	Oklahoma	3,504,917	1,490.9
District of Colombia	557,846	183.5	South Carolina	4,146,753	1,613.6
Florida	16,993,369	4,287.8	Tennessee	5,841,585	2,268.9
Georgia	8,746,849	3,003.7	Texas	22,099,136	12,369.8
Kentucky	4,116,780	1,877.2	Virginia	7,383,387	2,428.6
Louisiana	4,490,380	3,693.0	West Virginia	1,810,347	784.1
Maryland	5,512,477	1,550.5			
Total	**48,463,556**	**18,054.9**			
West					
Alaska	648,510	761.9	Nevada	2,241,700	654.2
Arizona	5,577,784	1,370.7	New Mexico	1,879,252	663.1
California	35,456,602	8,130.3	Oregon	3,562,681	1,049.2
Colorado	4,548,071	1,351.5	Utah	2,378,696	704.9
Hawaii	1,248,200	309.6	Washington	6,131,131	1,934.6
Idaho	1,368,111	466.6	Wyoming	501,915	461.2
Montana	917,885	375.9			
Total	**66,460,538**	**18,233.7**			
U.S. Total	**290,850,005**	**98,554.9**			

Source: Energy Information Administration (EIA), 2006, "State Energy Data System tables,"
http://www.eia.doe.gov/emeu/states/_states.html. Accessed October 23, 2006.

TABLE 6.21
U.S. Census Bureau's Regional Divisions of the United States

Northeast	
New England	Connecticut, Maine, Massachusetts, New Hampshire, Rhode Island, Vermont
Middle Atlantic	New Jersey, New York, Pennsylvania

South	
South Atlantic	Delaware, District of Colombia, Florida, Georgia, Maryland, North Carolina, South Carolina, Virginia, West Virginia
East South Central	Alabama, Kentucky, Mississippi, Tennessee
West South Central	Arkansas, Louisiana, Oklahoma, Texas

Midwest	
East North Central	Indiana, Illinois, Michigan, Ohio, Wisconsin
West North Central	Iowa, Kansas, Minnesota, Missouri, Nebraska, North Dakota, South Dakota

West	
Mountain	Arizona, Colorado, Idaho, Montana, Nevada, New Mexico, Utah, Wyoming
Pacific	Alaska, California, Hawaii, Oregon, Washington

Source: U.S. Census Bureau, 2006, "U.S. Regions," http://www.census.gov/geo/www/us_regdiv.pdf. Accessed October 23, 2006.

states, coal is often found closer to the surface and more easily extracted using such methods.

Electricity generation is another important component to energy use in the United States. Figure 6.18 demonstrates that coal accounted for 50 percent of the electricity produced in the United States in 2005. Natural gas and nuclear resources are the two other main sources used by the electric utility industry.

With the exception of hydroelectricity, renewable energy sources comprise a relatively small amount of the energy pie in the United States. However, since renewable energy technologies will become more important for future energy needs, two graphics depicting renewable energy consumption by type and by sector of end use are in figure 6.19.

Previous chapters discussed the importance of energy legislation and regulation to resource allocation for society. Tables 6.24–6.30 provide an overview of important U.S. energy legislation. Included in these selections are environmental and land use laws that have impacted energy industries in the United States.

FIGURE 6.13
U.S. Energy Consumption by Regional Division (2003)

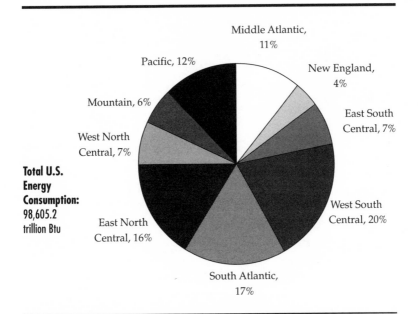

Middle Atlantic, 11%

Pacific, 12%

New England, 4%

Mountain, 6%

East South Central, 7%

West North Central, 7%

Total U.S. Energy Consumption: 98,605.2 trillion Btu

West South Central, 20%

East North Central, 16%

South Atlantic, 17%

Source: Energy Information Administration (EIA), 2006, "Total Energy Consumption Table," State Energy Data System, http://www.eia.doe.gov/emeu/states/_states.html. Accessed October 24, 2006.

TABLE 6.22
State Division by PAD District

PAD District	States
I	Maine, New Hampshire, Massachusetts, Rhode Island, Connecticut, Vermont, New York, Pennsylvania, New Jersey, Delaware, West Virginia, Maryland, Virginia, North Carolina, South Carolina, Georgia, Florida
II	Tennessee, Kentucky, Ohio, Michigan, Indiana, Illinois, Wisconsin, Missouri, Iowa, Minnesota, North Dakota, South Dakota, Nebraska, Kansas, Oklahoma
III	Alabama, Mississippi, Arkansas, Louisiana, Texas, New Mexico
IV	Montana, Wyoming, Colorado, Utah, Idaho
V	Arizona, California, Nevada, Oregon, Washington, Alaska, Hawaii

Source: Energy Information Administration (EIA), "Crude Oil Production," *Petroleum Supply Annual 2005,* Petroleum Navigator, http://tonto.eia.doe.gov/dnav/pet/pet_crd_crpdn_adc_mbbl_a.htm. Accessed October 24, 2006.

FIGURE 6.14
U.S. Crude Oil Production (1900–2005)

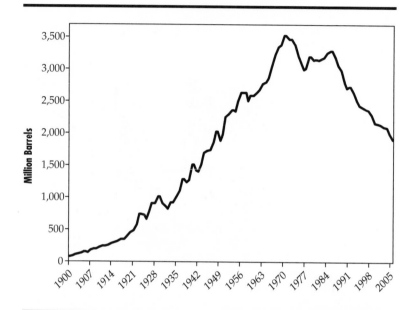

Source: Energy Information Administration (EIA), "Crude Oil Production," *Petroleum Supply Annual 2005,* Petroleum Navigator, http://tonto.eia.doe.gov/dnav/pet/pet_crd_crpdn_adc_mbbl_a.htm. Accessed October 24, 2006.

FIGURE 6.15
U.S. Crude Oil Production by PAD District (2005) (thousand barrels)

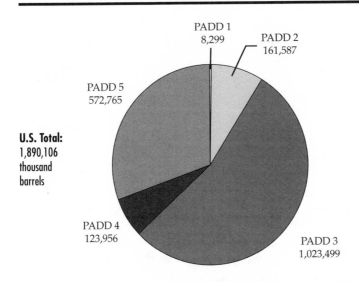

U.S. Total:
1,890,106
thousand
barrels

PADD 1
8,299

PADD 2
161,587

PADD 5
572,765

PADD 4
123,956

PADD 3
1,023,499

Source: Energy Information Administration (EIA), 2005, "Crude Oil Production," *Petroleum Supply Annual 2005,* Petroleum Navigator, http://tonto.eia.doe.gov/dnav/pet/pet_crd_crpdn_adc_mbbl_a.htm.

TABLE 6.23
U.S. Coal-Producing Regions

Region	States	Surface Mines (2005)	Underground Mines (2005)
Appalachian	Alabama, Georgia, eastern Kentucky, Maryland, North Carolina, Ohio, Pennsylvania, Tennessee, Virginia, and West Virginia	548	682
Interior Region (including Gulf Coast)	Arkansas, Illinois, Indiana, Iowa, Kansas, Louisiana, Michigan, Mississippi, Missouri, Oklahoma, Texas, and western Kentucky	34	72
Western Region	Alaska, Arizona, Colorado, Montana, New Mexico, North Dakota, Utah, Washington, and Wyoming	24	38

Source: Energy Information Administration (EIA), 2006, "Glossary of Terms," http://www.eia.doe.gov/glossary/index.html

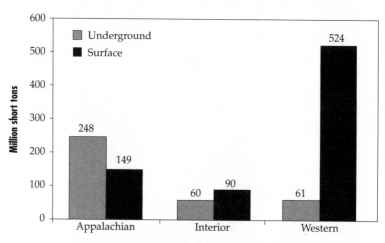

FIGURE 6.16
U.S. Coal Production by Coal-Producing Region (2005)

U.S. total coal production: 1,130 million short tons

Source: Energy Information Administration (EIA), 2005, *Annual Coal Report 2005,*
http://www.eia.doe.gov/cneaf/coal/page/acr/acr_sum.html. Accessed October 26, 2006.

FIGURE 6.17
U.S. Natural Gas Production (2004) (million cubic ft)

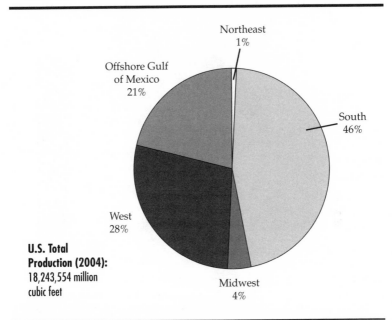

Northeast
1%

Offshore Gulf
of Mexico
21%

South
46%

West
28%

**U.S. Total
Production (2004):**
18,243,554 million
cubic feet

Midwest
4%

Source: Energy Information Administration (EIA), 2006, "Natural Gas Gross Withdrawals and Production," *Natural Gas Navigator,* http://tonto.eia.doe.gov/dnav/ng/ng_prod_sum_a_EPG0_FPD_mmcf_a.htm. Accessed October 27, 2006.

FIGURE 6.18
U.S. Net Electric Generation by Energy Source (2005)

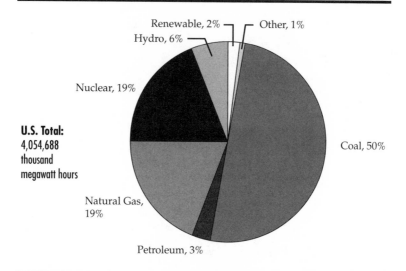

Renewable, 2% — — Other, 1%
Hydro, 6% —

Nuclear, 19%

U.S. Total:
4,054,688
thousand
megawatt hours

Coal, 50%

Natural Gas,
19%

Petroleum, 3%

Source: Energy Information Administration (EIA), 2006, "Net Generation by Energy Source by Type of Producer," *Electric Power Annual Report,* http://www.eia.doe.gov/cneaf/electricity/epa/epat1p1.html. Accessed October 27, 2006.

FIGURE 6.19

U.S. Renewable Energy Consumption (2004) (quadrillion Btu)

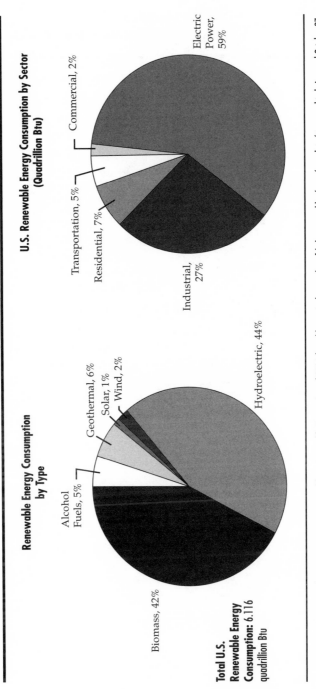

Renewable Energy Consumption by Type

Alcohol Fuels, 5%
Geothermal, 6%
Solar, 1%
Wind, 2%
Hydroelectric, 44%
Biomass, 42%

Total U.S. Renewable Energy Consumption: 6.116 quadrillion Btu

U.S. Renewable Energy Consumption by Sector (Quadrillion Btu)

Commercial, 2%
Transportation, 5%
Residential, 7%
Industrial, 27%
Electric Power, 59%

Source: Energy Information Administration (EIA), 2006, "Tables," *Renewable Energy Annual 2004,* http://www.eia.doe.gov/cneaf/solar.renewables/page/rea_data/rea_sum.html. Accessed October 27, 2006.

FIGURE 6.20
U.S. Petroleum Trade (1960–2005)

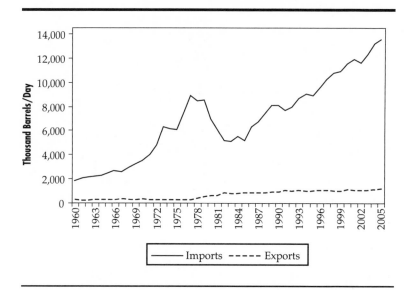

Source: Energy Information Administration (EIA), 2006, "Petroleum Overview, Selected Years, 1949-2005," *Annual Energy Review, 2005* (Washington DC: Department of Energy), 127. Accessed October 15, 2006. Accessed October 15, 2006.

FIGURE 6.21
Top Ten U.S. Petroleum Suppliers (2004)

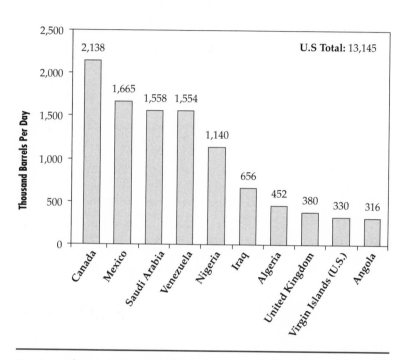

Source: Energy Information Administration (EIA), "Top Suppliers of U.S. Crude Oil and Petroleum, 2004," *Petroleum Supply Annual 2004,* http://www.eia.doe.gov/oil_gas/petroleum/info_glance/petroleum.html. Accessed October 1, 2006.

TABLE 6.24
National-Energy-Policy Legislation

Year	Legislation	Major points
1917	Food and Food Control (Lever) Act	Creates the U.S. Fuel Administration (USFA) to oversee operations of the coal industry; grants the administrative branch the power to fix coal prices. The USFA was dismantled in 1919.
1944	The Synthetic Liquid Fuels Act	Establishes a research program for the development of synthetic fuels as part of a strategic reserve initiative. The program was ended in 1954.
1950	Defense Production Act	Grants the secretary of the interior the authority to oversee power production and distribution for the Korean War effort.
1973	Emergency Petroleum Allocation Act	Establishes price controls for petroleum in response to the Arab Oil Embargo.
1974	Energy Supply and Environmental Coordination Act	Mandates that steps be taken to assess and coordinate the nation's energy resources with its energy needs; directs that coal resources be more effectively utilized; seeks to adjust environmental regulations to meet energy needs.
1975	Energy Policy and Conservation Act	Addresses supply issues and develops conservation programs for managing energy demand; mandates the development of a strategic petroleum reserve; seeks to improve energy efficiency of the nation's energy industries.
1977	Energy Reorganization Act	Consolidates U.S. energy agencies with the creation of the Department of Energy (DOE).
1978	National Energy Conservation Policy Act	Mandates the federal government to explore energy-efficient technologies in buildings; promotes the use of renewable energy technologies in public buildings.
1978	Energy Tax Act	Provides tax incentives for households to implement energy conservation measures.
1980	Windfall Profits Tax Act	Issues an excise tax on domestic oil companies. The tax was levied on the revenue difference between the market price and a government determined base price of oil.
1982	Energy Emergency Preparedness Act	Updates the Energy Policy and Conservation Act with a focus on further development of strategic petroleum reserves; includes measures on how the federal government will respond to energy-emergencies and fuel shortages.
1992	Comprehensive National Energy Policy Act	Mandates energy efficiency standards for federal buildings and provides incentives to states, businesses, and households to adopt energy-efficient technology.
2005	Energy Policy Act	Provides incentives for the development of renewable energy sources, most notably to the biofuels industry to stimulate the use of ethanol-burning vehicles; provides tax and other incentives for the development of nuclear energy resources; continues to grant subsidies for oil and gas drilling and exploration activities.

TABLE 6.25
Nuclear-Energy Legislation

Year	Legislation	Major points
1946	Atomic Energy Act	Establishes the Atomic Energy Commission (AEC) and the Joint Commission on Atomic Energy (JACE) as oversight agencies for nuclear development; mandates that all nuclear resources be owned by the federal government.
1954	Atomic Energy Act	Allows for private ownership of nuclear power as a stimulus for industry to develop nuclear power resources.
1957	Price-Anderson Act	Further promotes nuclear development in the United States by limiting liability for potential nuclear accidents.
1974	Energy Reorganization Act	Creates the Energy Research and Development Administration (ERDA) to take over duties of the AEC.
1980	Nuclear Safety Research, Development, and Demonstration Act	Mandates the development of safety standards for the construction and operation of nuclear power plants.
1982	Nuclear Waste Policy Act	Mandates that two sites, one in the East and one in the West, be established as nuclear waste repositories for waste generated in the United States.
1987	Nuclear Waste Policy Act	Revises the 1982 act to mandate that one central storage facility be developed and recommended Yucca Mountain as the site for the repository.
1988	Price-Anderson Amendments	Raises the liability limits that the federal government would pay plant owners/operators in the event of an accident.

TABLE 6.26
Renewable-Energy Legislation

Year	Legislation	Major points
1970	Geothermal Steam Act	Authorizes federal development of geothermal resources and allows the Department of the Interior to lease land for geothermal development.
1974	Solar Energy Research, Development, and Demonstration Act	Authorizes federal support of research and development of solar energy.
1974	Solar Heating and Cooling Demonstration Act	Mandates the demonstration of solar technology for heating and cooling of residential dwellings and commercial structures.
1978	Solar Photovoltaic Energy Research, Development, and Demonstration Act	Mandates that the secretary of energy spearhead efforts to implement vigorous research and development programs for improving the efficiency and reducing the cost of solar photovoltaic technology.
1980	Energy Security Act	Designed to develop solutions for energy security issues; mandates research and development funding for solar, geothermal, ocean thermal, biomass, and other renewable technologies.

TABLE 6.27
Regulation of Electricity and Utilities

Year	Legislation	Major points
1920	Water Power Act	Allows for federal oversight of hydroelectric power generation on navigable streams; establishes the Federal Power Commission (later became the Federal Energy Regulatory Commission [FERC]).
1933	Tennessee Valley Authority Act	Provides federal funding for a multipurpose river project; establishes publicly owned hydroelectricity resources and flood control network in the Tennessee Valley.
1935	Federal Power Act	Allows the government to regulate and oversee utility rates.
1935	Public Utility Holding Company Act (PUHCA)	Abolishes utility holding companies and regulates utility transactions.
1936	Rural Electrification Act	Seeks to increase electricity distribution to rural areas by providing financial support and incentives to nonprofit cooperatives for establishing electric utilities in rural areas.
1938	Natural Gas Act	Gives the Federal Power Commission authority to regulate interstate sales of natural gas.
1968	Natural Gas Pipeline Safety Act	Mandates that safety standards be developed for the transport of natural gas and other fuels by pipeline.
1978	Natural Gas Policy Act	Removes the distinction between inter- and intrastate natural gas markets by establishing wellhead pricing and gives the federal government a larger role in natural gas pricing.
1978	Public Utilities Regulatory Policies Act (PURPA)	Requires utility companies to purchase electricity produced by cogeneration.
2005	Energy Policy Act	Removed PURPA regulations and nullified PUHCA in an effort to promote industry deregulation.

TABLE 6.28
Pollution-Control Acts

Year	Legislation	Major points
1924	Oil Pollution Act	Mandates pollution control measures for the practices of drilling, pumping, refining, and transporting petroleum. The act marks the first time that safety and environmental standards for oil production are required by the federal government. Although largely developed by oil-friendly interests, the law recognizes environmental damage that occurs from oil production.
1972	Clean Water Act	Establishes a permitting system for the discharge of waste into the nation's navigable waterways and requires that pollution control mechanisms be installed to remove pollutants from water prior to discharge.
1980	Resource Conservation and Recovery Act (RCRA)	Mandates standards and procedures for the disposal of hazardous waste. This law affected energy industries and their practices of handling and disposal of wastes accrued during extraction and refining processes.
1980	Comprehensive Environmental Response, Compensation, and Liability Act	Develops a regulatory structure for cleaning up abandoned hazardous waste sites. Many of the contaminated, or Superfund, sites were owned and operated by the DOE as facilities used in energy resources testing and development.
1990	Oil Pollution Act of 1990	Eliminates federal liability caps for accidents resulting from negligence; requires that all tankers operating in U.S. waters demonstrate financial responsibility; mandates that single-hull tankers be phased out.

TABLE 6.29
Clean Air Acts

Year	Legislation	Major points
1955	Clean Air Act	Recognizes air pollution as a local problem and restricts federal involvement to technical assistance with pollution-abatement programs.
1963	Clean Air Act	Grants the federal government authority to intervene in interstate air-pollution matters at the request of state governments.
1967	Clean Air Act	Mandates the establishment of Air Quality Control Regions that require states to develop air quality standards.
1970	Clean Air Act	Develops National Ambient Air Quality Standards (NAAQS) and requires states to develop implementation plans to meet NAAQS.
1977	Clean Air Act Amendments	Develops specific classes of air quality control regions based on pollution severity and proximity to conservation areas; permits the enforcement of federal air quality control measures if state implementation plans are not effective.
1990	Clean Air Act Amendments	Requires more stringent standards for six types of air pollutants (see chapter 3); establishes a permitting program for emission sources; establishes cap-and-trade mechanisms for pollution reduction.
2004	Clear Skies Act	Revises deadlines for pollution caps on NO_x and SO_2 emissions.

TABLE 6.30
Federal Lands Acts

Year	Legislation	Major points
1920	Mineral Leasing Act	Allows the lease of federal land to energy companies for oil and gas development.
1953	Outer Continental Shelf Lands Act	Establishes the Department of the Interior (DOI) as the main leasing agent of offshore areas for oil and gas drilling.
1976	Federal Coal Leasing Amendments Act	Establishes provisions for the leasing of federal lands for coal mining.
1976	Federal Land Policy and Management Act	Mandates the DOI to coordinate land use and the development of environmental impact statements (EIS) among the different land-management agencies. The act's environmental stipulations also have implications for the energy industry.
1977	Surface Mining Control and Reclamation Act	Mandates that coal companies reclaim and restore land after surface-mining operations cease; creates the Office of Surface Mining to enforce the legislation.

References

Marland, G., T. A. Boden, and R. J. Andres. 2006. "Global, Regional, and National CO$_2$ Emissions." In *Trends: A Compendium of Data on Global Change*. Carbon Dioxide Information Analysis Center, Oak Ridge

National Laboratory, TN. U.S. Department of Energy. http://cdiac.ornl .gov/trends/emis/tre_glob.htm (accessed October 1, 2006).

Silvi, C. 2004. "Solar Energy." In *2004 Survey of Energy Resources*, World Energy Council, 298. 20th ed. Amsterdam: Elsevier.

Varming, S. "Wind Energy." In *2004 Survey of Energy Resources*, World Energy Council, 369. 20th ed. Amsterdam: Elsevier.

7

Directory of Organizations

Introduction

This chapter provides a directory of organizations, agencies, and associations that operate in the broad area of energy use in society. Because energy is so closely tied with the environment and human development, the organizations represented in this chapter provide a glimpse into the complex aspects of energy use worldwide. They include professional organizations, nonprofit and nongovernmental associations, and government agencies. They specialize in areas of individual fossil fuels, sustainable development, climate change, overall energy dynamics, and environmental issues. The goal of this chapter is to provide an annotated quick-reference guide to the large number of organizations operating in all areas of energy dynamics.

African Energy Policy Research Network/Foundation for Woodstove Dissemination (AFREPREN/FWD)
http://www.afrepren.org/

The AFREPREN/FWD is a nongovernmental organization that initiates energy policy research initiatives primarily in Eastern and Southern Africa. They promote collaborations between energy researchers and policymakers in the areas of energy reform, energy service provision for the urban poor, renewable energy in rural development, and overall energy trends and investments in the African energy sector. Their Web site provides an overview of research objectives, a list of publications from the AFREPREN/FWD, and updated news and events.

African Wind Energy Association (AfriWEA)
http://www.afriwea.org/

Formed in 2002, the African Wind Energy Association is a nonprofit organization that promotes wind energy development on the African continent. It supports the growth of the wind energy industry by providing a network of political and technical support, promoting cooperation and collaboration within the industry, and facilitating communication among African wind energy member committees. The AfriWEA Web site offers information on wind projects in member countries, financing opportunities, the development of "mini-grids," and links to companies that provide secondhand wind turbines.

Alliance to Save Energy (ASE)
http://www.ase.org/

The Alliance to Save Energy is a nonprofit advocacy organization that promotes the implementation of energy efficiency measures within the existing market structure. It is involved in research, educational programs, energy efficiency projects, and technology development. The ASE Web site contains information about efficiency measures taken in a variety of energy products and sectors as well as provides overviews of its activities in over twenty-six countries.

American Association of Petroleum Geologists (AAPG)
http://www.aapg.org/

Founded in 1917, this professional organization for geoscientists lists members in over 116 countries. Its main goals are to foster geological research, promote the advancement of technology, and serve as an information organization for geoscientists specializing in petroleum resources. The AAPG Web site offers energy facts, links to meetings and events, information about publications, and overviews of AAPG grants and programs.

American Coal Ash Association (ACAA)
http://www.acaa-usa.org/

The American Coal Ash Association is a not-for-profit organization that seeks to advance the use of coal combustion products (CCP), such as fly ash, for commercial purposes. It provides educational workshops, technical materials, and information to advance the use of CCP materials. The ACAA Web site offers background information on CCPs, a library of resources, an overview of regulatory issues, and links to energy companies and coal ash industries.

American Coal Council (ACC)
http://www.americancoalcouncil.org/

The American Coal Council seeks to promote the development and use of coal as an energy resource. It serves as an educational resource, networking forum, and information source for the U.S. coal industry. The ACC Web site features links to industry suppliers and industry statistics that are updated daily.

American Council for an Energy Efficient Economy (ACEEE)
http://www.aceee.org/

The American Council for an Energy Efficient Economy is a nonprofit organization that promotes the inclusion of energy efficiency measures in business and policy. It collaborates with industry, government, and academic institutions to educate consumers about energy efficient choices, provide policy assessments, and organize workshops. The ACEEE Web site offers an overview of program areas and provides information for energy consumers.

American Hydrogen Association (AHA)
http://www.clean-air.org/

The American Hydrogen Association is a nonprofit organization dedicated to advancing the use of hydrogen energy systems. It provides promotional and educational materials and products and reports on current technical, economic, political, and social issues associated with hydrogen technology through the publication of a quarterly newsletter. The AHA Web site features current news associated with the use of hydrogen technology in business and provides information on how to construct personal hydrogen energy systems.

American Institute of Mining, Metallurgical, and Petroleum Engineers (AIME)
http://www.aimeny.org/

The American Institute of Mining, Metallurgical, and Petroleum Engineers is a nonprofit organization that represents professionals in the engineering sciences associated with the production of minerals, metals, energy sources, and materials. Established in 1871, it is considered one of the founding engineering societies in the United States. The AIME Web site provides an overview of the organization and its units, information about meetings and events, a library of resources, and announcements for awards and scholarships.

American Oil and Gas Historical Society (AOGHS)
http://www.aoghs.org/

The American Oil and Gas Historical Society is a nonprofit organization that seeks to preserve the history of the U.S. oil and natural gas industry through education, advocacy, and historical material preservation. It promotes energy education, networks with museums and historical societies, and provides a communication network for oil and natural gas museums. The AOGHS Web site provides museum links, access to the society's quarterly newsletter, articles and historic photos, and oil field images.

American Petroleum Institute (API)
http://www.api.org/

The American Petroleum Institute is a trade association representing all segments of the oil and natural gas industry, including producers, refiners, suppliers, pipeline operators, marine transporters, and service and supply companies. It acts as a lobbying organization for the petroleum industry as well as an information organization, providing data and statistics, standards for operation, and certification. The API Web site provides information about oil and natural gas, data and statistics for the petroleum industry, an overview of policy issues affecting the industry, and information about trainings, API standards, and certification programs.

American Public Power Association (APPA)
http://www.appanet.org/

The American Public Power Association is an organization representing community-owned electric utility operators in the United States. Established in 1940, it seeks to promote the provision of reliable electricity services by supporting the public policy interests of its members. The APPA Web site offers information about regulatory issues, industry standards, and announcements for professional development events that are pertinent to the electric utility industry.

American Solar Energy Society (ASES)
http://www.ases.org/

The American Solar Energy Society is a nonprofit organization that promotes the development of sustainable energy for the United States. It seeks to advance the development of solar and other renewable technologies through outreach, education, and collaboration initiatives. The ASES Web site contains information about programs, projects, and important events and provides general information about renewable energy.

American Wind Energy Association (AWEA)
http://www.awea.org/

The American Wind Energy Association is the national trade association for the wind energy industry in the United States. It represents the interests of wind energy professionals to policymakers, provides information about wind projects and companies worldwide, and promotes the development of wind technology. The AWEA Web site offers up-to-date news, educational materials, legislative updates, research information, and technical assistance for the wind energy industry.

Appliance Standards Awareness Project (ASAP)
http://www.standardsasap.org/

Founded in 1999, Appliance Standards Awareness Project is a committee of environmental and energy professionals that seeks to gain support for the adoption of appliance and equipment efficiency standards. It provides technical support for governments and industries that are interested in advancing state standards for appliance and instrument efficiency. The ASAP Web site features updates about ongoing initiatives at the federal and state levels to advance energy efficiency standards.

Asian Regional Research Programme in Energy, Environment and Climate (ARRPEEC)
http://www.arrpeec.ait.ac.th/

The Asian Regional Research Programme in Energy, Environment and Climate is a network of Asian research institutes that examine regional issues in the areas of energy, environment, and climate. The institutes seek to link research initiatives in participating countries for the purpose of streamlining efforts to reduce greenhouse gas emissions. The ARRPEEC Web site provides an overview of the network's three-phase research project, news and events related to climate issues in the Asian region, and links to publications.

Association for the Study of Peak Oil and Gas (ASPO)
http://www.peakoil.net/

The Association for the Study of Peak Oil and Gas is made up of scientists concerned about the impact of oil and natural gas depletion. Members conduct research in the area of oil and gas reserves, depletion modeling, and raising awareness of the issue of resource depletion. Formed in Germany in 2000, the ASPO currently represents members in fourteen countries. The ASPO Web site offers up-to-date commentary and information about the issue of peak oil.

Association of Energy and Environmental Real Estate Professionals (AEEREP)
http://www.aeerep.org/

The Association of Energy and Environmental Real Estate Professionals is a nonprofit organization that seeks to educate real estate professionals and consumers about choices in energy savings and environmental stewardship in the purchase of property or real estate. Its goal is to advance the green market industry in home design and products through advocacy and education. The AEEREP Web site provides information about association programs and links to similar organizations.

Association of Energy Engineers (AEE)
http://www.aeecenter.org/

The Association of Energy Engineers is a professional organization for engineers involved in the areas of utility regulation and deregulation, plant engineering, energy efficiency, facility management, and environmental compliance. It provides technical support, certification programs, conferences, and information resources to members. The AEE Web site provides information for engineering professionals, industry newsletters, and access to professional journals for members.

Association of Energy Service Companies (AESC)
http://www.aesc.net/

Established in 1956, the Association of Energy Service Companies operates as a trade organization to represent the interests of energy service companies. Its programs provide information and assistance to members in the areas of safety, technology development, government regulations, and industry training. The AESC Web site provides access to safety procedures and guidelines, a directory of companies that provide products and equipment for the oil and gas industry, and access to AESC's quarterly newsletter, *Field Reports.*

Association of Energy Services Professionals (AESP)
http://www.aesp.org/

The Association of Energy Services Professionals is a nonprofit professional organization representing the interests of energy service providers, energy marketers, utility companies, and end users of energy services. It promotes education, sponsors conferences and training, and provides networking opportunities for industry participants. The AESP Web site offers information on upcoming events, a career center, and an association newsletter.

Austrian Energy Agency
http://www.eva.ac.at

The Austrian Energy Agency is a nonprofit organization that focuses on energy and policy research. Formed in 1977, it works with the Austrian government to analyze technical issues, develop long-term energy strategies, review energy policy, and provide information and scientific support for energy activities on the national and international levels. The agency's Web site provides an overview of energy projects, energy data, and publications.

California Energy Commission
http://www.energy.ca.gov/

The California Energy Commission is the primary energy agency in the state of California. It is responsible for the planning and implementation of California's energy polices, licensing the state's thermal power plants, and developing energy technologies that support renewable energy and ensure affordable and reliable energy for California. The commission's Web site provides an overview of various state programs and regulations and has links to divisions within the agency.

Canadian Association of Petroleum Producers (CAPP)
http://www.capp.ca/

The Canadian Association of Petroleum Producers is a trade and advocacy organization for the Canadian oil and natural gas industry. It promotes public policy in support of the industry, facilitates information sharing, develops codes of practice, and serves as a networking forum for industry leaders. The CAPP Web site offers information about the Canadian oil and natural gas industry, an overview of relevant issues concerning each province, and an overview of the industry's environmental stewardship initiatives.

Center for Applied Energy Research
http://www.caer.uky.edu/

An affiliate of the University of Kentucky, the Center for Applied Energy Research is a research and development organization that provides applied research opportunities for energy industries, with a particular emphasis on the coal industry in the United States. The center's Web site provides an overview of current and past research and includes links for teachers, students, and professionals to sites focusing on coal resources in Kentucky and the United States.

Center for Energy and Economic Development (CEED)
http://www.ceednet.org/ceed/

The Center for Energy and Economic Development is a nonprofit organization that promotes the advancement of coal-based electricity. It seeks to educate the public and the government about the benefits of coal technologies and promotes the enhancement of clean coal technologies. The CEED Web site provides information about key issues that impact the coal utilities (e.g., mercury, climate change, etc.) as well as about environmental progress and clean coal technology, and offers state profiles on electricity generation.

Center for Resource Solutions (CRS)
http://www.resource-solutions.org

The Center for Resource Solutions is a nonprofit organization that seeks to increase the supply and demand for renewable sources of energy. It promotes the development of a renewable energy market by implementing programs that examine renewable energy policy and regulation. It provides a networking arena for businesses and government agencies to facilitate growth in the renewable energy industry. The CRS Web site includes information about measurement and verification strategies and clean energy policy and offers technical assistance to utility companies, energy developers, state energy offices, and private energy companies.

China Energy Group
http://china.lbl.gov/china.html

An affiliate of the Lawrence Berkeley National Lab, the China Energy Group is a research collaborative that seeks to understand Chinese energy dynamics, promote energy efficiency in Chinese institutions, and examine the role that energy plays in current and future Chinese society. The group's Web site offers information about energy use in Chinese buildings, an overview of renewable resources, and an analysis of Chinese energy policy.

Chinese Renewable Energy Industries Association (CREIA)
http://www.creia.net/cms_eng/_code/english/

The Chinese Renewable Energy Industries Association is an organization that seeks to address environmental issues associated with polluting energy industries in China by promoting the development and use of renewable energy technologies. It provides a collaborative bridge between regulatory agencies and industry and promotes renewable research and development activities. The CREIA Web site provides an overview of the renewable energy industry in China, Chinese energy policies and regulations, and association programs and projects.

Climate Action Network (CAN)
http://www.climatenetwork.org/

The Climate Action Network is a group of nongovernmental organizations that seeks to mitigate human-induced climate change through government and individual action. It provides an arena for information exchange and strategic coordination among members. It promotes a three-tiered approach to addressing climate change, involving the Kyoto Agreement, a decarbonization approach, and adaptation. The CAN Web site provides resources for members, overviews of climate policies, and general information and news regarding the issue of global climate change.

Climate Institute
http://www.climate.org

The Climate Institute is a nonprofit organization that provides information on climate science and policy to decision makers, sponsors conferences and symposia on climate research, and promotes practical approaches for achieving a global climate balance. The Institute's Web site provides information about research that is being done in the area of climate science, offers updated information about the effects of global climate change, and describes what individuals can do to address the issue of climate change.

Coal Utilization Research Council (CURC)
http://www.coal.org/

Formed in 1997, the Coal Utilization Research Council works to build collaborations between the coal industry and government to promote the research and development of clean coal and other coal-related technologies. The CURC Web site provides basic information on coal and clean coal technologies, a summary of CURC goals, activities, and successes, and a list of upcoming U.S. legislation that impacts the use of coal resources.

Coalition for Affordable and Reliable Energy (CARE)
http://www.careenergy.com

Formed in 2000, the Coalition for Affordable and Reliable Energy seeks to ensure the availability of affordable and reliable sources of energy for U.S. society, with a particular focus on clean coal technologies. It promotes the increased use of domestic resources, policies that diversify fuel resources, and the development of advanced energy technologies. The CARE Web site provides technology updates for the coal and utility industries, information about electricity supply in the United States, and an overview of environmental initiatives that have allowed for the cleaner use of coal resources.

Convergence Research
http://www.converger.com/

Convergence Research is a technical and policy consulting group for energy, water, and transportation industries. It provides policy analysis, information, and technology tools for public, private, and nonprofit groups. The group's Web site offers energy policy analyses, reference databases, and research publications.

Cooperative Research Center for Coal in Sustainable Development (CCSD)
http://www.ccsd.biz/

The Cooperative Research Center for Coal in Sustainable Development is a research organization that brings together experts in coal research and sustainability issues for the purpose of promoting the sustainable use of coal as an energy source for future generations. The CCSD Web site provides overviews of the various programs the organization supports as well as information and reference materials.

Domestic Petroleum Council (DPC)
http://www.dpcusa.org/

The Domestic Petroleum Council is a trade organization for independent oil and natural gas exploration and production (E&P) companies. It promotes the development of public policies that encourage the responsible development of oil and natural gas. The DPC Web site offers overviews of recent E&P technologies, information on environmental stewardship, an overview of issues affecting the natural gas industry, and policy assessments provided in the format of downloadable reports.

Edison Electric Institute (EEI)
http://www.eei.org/

The Edison Electric Institute represents shareholder-owned electric companies. Established in 1933, the EEI advocates regulatory and legislative policy on behalf of its members and provides information and business resources. The EEI Web site provides an overview of energy infrastructure, environmental issues, electricity policy, and similar topics, as well as links to publications for members.

Electric Power Research Institute (EPRI)
http://my.epri.com

Established in 1973, the Electric Power Research Institute is a nonprofit research organization of scientists and engineers working to develop solutions to issues involving the electric power industry. The EPRI Web site provides information and research on a number of issues, including the environment, electricity generation, power delivery, and nuclear issues.

Energistics
http://www.energistics.org

Energistics is a not-for-profit organization devoted to uniting and serving professionals in the petroleum industry. It facilitates exploration and production information-sharing, promotes business process integration, and uses collaboration as a tool for developing business solutions to industry problems. The Energistics Web site has a standards resource center where users can search or browse standards pertinent to their business; a discussion forum where professionals can post concerns, questions, and other comments; and news and information about upcoming events relevant to the petroleum industry.

Energy Action Coalition
http://www.energyaction.net/main/

The Energy Action Coalition is a network of more than thirty organizations working to strengthen the involvement of youth in clean energy advocacy. Focusing in the strategic areas of campuses, communities, corporate practices, and politics, Energy Action provides a platform for communicating about various campaigns that seek to promote the use of sustainable energy technologies. The coalition's Web site provides links to a variety of student and professional organizations and information about current and past campaigns.

Energy Advocates
http://www.energyadvocates.org/

Established in 1974, Energy Advocates seeks to raise public awareness about important energy issues. The organization promotes the image of the energy industry while enhancing public understanding of the importance of energy to society. The Web site provides information sources on energy, news updates, and energy factoids.

Energy and Environmental Building Association (EEBA)
http://www.eeba.org

The Energy and Environmental Building Association was formed in 1982 for the purpose of developing standards for energy efficient building construction. It provides educational materials and workshops aimed at enhancing energy efficient building construction in communities. The EEBA Web site offers an overview of building technology, educational materials, and a link to the EEBA Institute, which offers courses in energy efficient building design and construction.

Energy and Mineral Law Foundation (EMLF)
http://www.emlf.org/

The Energy and Mineral Law Foundation is a nonprofit organization that provides information on legal issues associated with energy and mineral industries. It provides information resources and an educational forum for industry, government, legal scholars, and attorneys in the area of natural resources law. The EMLF Web site offers access to peer-reviewed papers published in the *Annual Institutes of the Energy and Mineral Law Foundation.*

Energy and Resources Institute, The (TERI)
http://www.teriin.org/

Established in 1974, the Energy and Resources Institute works to promote sustainable development for Indian society. It focuses on the utilization of natural and human resources in a way that is both equitable and environmentally sound. The TERI Web site offers information on energy and sustainable development projects within Indian society.

Energy Efficiency and Conservation Authority (EECA)
http://www.eeca.govt.nz

The Energy Efficiency and Conservation Authority is a New Zealand government organization established to implement the Energy Efficiency and Conservation Act. It seeks to raise awareness about energy efficiency and help businesses and communities implement energy conservation programs. The authority's Web site contains information about how energy efficiency can be achieved in various energy sectors.

Energy Foundation–China (EF–China)
http://www.ef.org

Energy Foundation–China implements the China Sustainable Energy Program, which furthers the use of renewable energy and energy efficiency in Chinese energy sectors. The program promotes capacity building to enhance energy savings in the Chinese economy. It accomplishes this task by working collaboratively with agencies, energy experts, and nongovernmental organizations. The EF–China Web site features grant opportunities, guidelines for grant applications, and an overview of programs that have been funded by the Energy Foundation.

Energy Information Administration (EIA)
http://www.eia.doe.gov/

The Energy Information Administration is an agency of the U.S. Department of Energy. It provides up-to-date national and international energy statistics, reports, and energy analyses. The EIA Web site offers up-to-date statistics for all U.S. states and territories as well as for all countries and regions worldwide. The site also provides monthly and annual energy reports for each energy resource, chronologies of energy events, links to general information concerning energy topics, and an interactive page for children on energy basics.

Energy Justice Network
http://www.energyjustice.org/

The Energy Justice Network is a grassroots organization providing support to communities that have experienced adverse effects from energy and waste industries. It focuses on issues of environmental justice in low-income and minority populations and promotes completely replacing nuclear and fossil fuels with renewable energy technologies within the next twenty years. The network's Web site has information about energy systems, summaries of alternative fuels and biomass incineration technologies, solutions that promote conservation and efficiency, and links to energy youth groups and action coalitions.

Environmental Energy Technologies Division (EETD)
http://eetd.lbl.gov

The Environmental Energy Technologies Division of the Ernst Orlando Lawrence Berkeley National Laboratory performs research and development on sustainable and environmentally friendly energy technologies. Its goals are to improve energy technologies, conduct energy analyses in both industrialized and developing countries, and examine the relation between environment and energy use. The EETD Web site provides extensive information about energy efficiency and environmental research.

European Energy Network (ENR)
http://www.enr-network.org/

The European Energy Network is an association of European organizations involved in the development, management, and provision of energy programs in the areas of renewable energy and energy efficiency. It facilitates communication and collaboration between member organizations and provides technical support to the European energy community. The ENR Web site serves as an information exchange for member organizations and includes summaries of ongoing and completed projects.

European Oil and Gas Innovation Forum (EUROGIF)
http://www.eurogif.org/

The European Oil and Gas Innovation Forum is a trade organization representing the interests of its membership. It was formed in 1996 by major European companies and national industry associations involved in oil and gas manufacturing and service and supply industries. The EUROGIF Web site provides technical and industry resources.

European Pipeline Research Group (EPRG)
http://www.eprg.net/

Made up of pipe manufacturers and gas transmission companies, the European Pipeline Research Group conducts research in gas-pipeline safety. It was formed in 1972 to study the issue of fractures found in gas transmission pipelines. The EPRG Web site provides results of pipeline research efforts organized into the three technical areas of corrosion, design, and materials.

Federal Energy Regulatory Commission (FERC)
http://www.ferc.gov/

The Federal Energy Regulatory Commission is the federal agency responsible for oversight of U.S. energy industries. FERC examines economic, environmental, and safety issues involved in the interstate transmission of oil, natural gas, and electricity. It is also responsible for the regulation of natural gas and hydropower projects. The FERC Web site contains information on energy regulation within the United States, an overview of energy industries, legal resources, and congressional and regulatory updates.

Gas Technology Institute (GTI)
http://www.gastechnology.org

A research, development, and training organization for the natural gas industry, the Gas Technology Institute develops and promotes technology-based solutions for energy challenges. The GTI Web site provides career and professional development information with links to trainings, conferences, and commercial opportunities for companies to employ recently patented material. It also contains information on software products and services for the industry.

Geothermal Resources Council (GRC)
http://www.geothermal.org/

Formed in 1970, the Geothermal Resources Council is a nonprofit educational organization for professionals in the geothermal industry. It encourages the development of geothermal resources and facilitates the transfer of geothermal technology and information. The GRC Web site provides news and articles about geothermal energy, contact information for corporate and independent members of the international geothermal community, and general information about geothermal energy.

Global Wind Energy Council (GWEC)
http://www.gwec.net/

Established in 2005, the Global Wind Energy Council is an international forum for the wind energy industry. Its mission is to promote the development and use of wind power as a leading energy resource. To accomplish this mission, the GWEC works in areas of policy development, business leadership, global outreach, and education. The GWEC Web site provides information about wind energy industries worldwide, offers information about upcoming events, and provides links to publications about economics and recent technology in the wind industry.

Independent Petroleum Association of America (IPAA)
http://www.ipaa.org/

The Independent Petroleum Association of America is a trade organization for independent exploration and production companies operating in the U.S. It provides technical and statistical information about the industry for its members. The IPAA Web site features a business center that provides information on financing institutions for the industry and insurance providers, as well as an industry calendar that highlights important meetings and events. It also provides an overview of issues impacting independent producers of oil and natural gas.

Innovation Center for Energy and Transportation (ICET)
http://www.icet.org.cn/

Formed in 2006, the Innovation Center for Energy and Transportation is a Chinese-based nongovernmental organization that seeks to promote fuel efficiency in automobiles, reduce greenhouse gases from the transportation sector, and raise awareness about the importance of clean vehicle technologies. The ICET Web site provides access to information about biofuels, green vehicles, support for associated businesses, and fuel efficiency standards and current renewable energy law in China.

Institute for Energy and Sustainable Development (IESD)
http://www.iesd.dmu.ac.uk/

The Institute for Energy and Sustainable Development is a research and consulting school that offers instruction in the area of sustainable development. It offers programs in the area of energy, sustainable building, climate change, and sustainable development. The IESD Web site provides an overview of degree programs and consulting activities.

Intergovernmental Panel on Climate Change (IPCC)
http://www.ipcc.ch/

The Intergovernmental Panel on Climate Change was established to collate scientific, technical, and socioeconomic research in the area of global climate change. The goals of the IPCC are to understand the potential impacts of human-induced climate change through the assessment of peer-reviewed, published research. The IPCC Web site provides downloadable versions of assessment reports and other publications, information about committees and projects, graphics from reports, and recent press releases and event information.

International Association for Hydrogen Energy (IAHE)
http://www.iahe.org

The International Association for Hydrogen Energy is a collaborative professional organization that strives to promote the use and dissemination of hydrogen technology as a primary energy resource for mankind. It sponsors international workshops, publishes the *International Journal of Hydrogen Energy,* and provides general information to the public about hydrogen resources. The IAHE Web site offers conference and publication information as well as links to organizations associated with the hydrogen industry.

International Atomic Energy Agency (IAEA)
http://www.iaea.org/

The International Atomic Energy Agency was established in 1957 as an international nuclear watchdog organization. It provides technical support for the development of peaceful nuclear programs, ensures the safety and security of nuclear facilities, and verifies that nuclear programs worldwide are not used for military purposes. The IAEA Web site provides access to information about its programs and activities, a database of nuclear energy information, and nuclear standards, development reports, and technical publications.

International Committee for Coal and Organic Petrology (ICCOP)
http://www.iccop.org/

Formed in 1951, the International Committee for Coal and Organic Petrology is a professional organization for scientists specializing in the area of coal chemistry and composition. It provides standards, training, and accreditation to laboratories studying coal science. The ICCOP Web site provides information about professional meetings and summaries of working committees.

International Council for Local Environmental Initiatives (ICLEI)
http://www.iclei.org/

Formed in 1990, the International Council for Local Environmental Initiatives is an international organization representing local governments that are interested in implementing sustainable development measures within their communities. It offers technical consulting, training, and information for member governments and organizations. The ICLEI Web site provides information about council programs and offices located around the world.

International Energy Agency (IEA)
http://www.iea.org/

The International Energy Agency is an international energy organization that serves as an advising body for twenty-six member countries. Its efforts are focused in three main areas: energy security, economic development, and environmental sustainability. The IEA Web site provides policy analyses, publications, and statistical data on energy use worldwide.

International Institute for Energy Conservation (IIEC)
http://www.iiec.org/

Founded in 1984, the International Institute for Energy Conservation is a nonprofit, nongovernmental organization that works to promote the sustainable use of energy, water, and land resources. It provides technical and institutional support, constructs policies, and implements programs that incorporate efficiency measures in the design of energy systems. The IIEC Web site contains information on services provided by the IIEC, an e-newsletter, news, and other information in the area of energy efficiency.

International Rivers Network (IRN)
http://www.irn.org/

The International Rivers Network is an activist network that works to discourage the development of large dams for the purposes of power generation, irrigation, and flood control. In collaboration with communities, social movements, and nongovernmental organizations, it advocates the development of alternative energy resources and water projects, conducts research in the area of sustainable water and energy solutions, and works to educate communities and development organizations about the adverse impact of large dams. The IRN Web site offers extensive information about current and past dam projects, an overview of water and energy alternatives, and provides a venue for citizen action.

International Solar Energy Society (ISES)
http://www.ises.org

Founded in 1954, the International Solar Energy Society is a nonprofit nongovernmental organization that promotes the growth of renewable energy technology. It works to promote research and development initiatives in renewable energy, facilitates the transfer of technology, provides a forum for the global renewable energy community, and promotes the dissemination of renewable energy information. The ISES Web site provides information about upcoming conferences and events as well as links to renewable energy publications.

Interstate Oil and Gas Compact Commission (IOGCC)
http://www.iogcc.state.ok.us/

The Interstate Oil and Gas Compact Commission is a multistate government agency that seeks to advance the development of domestic energy resources. It acts as a voice for participating state governments in Congress. The IOGCC Web site provides updated information on federal and state energy regulations and links to information and events about energy regulatory concerns.

Korean Peninsula Energy Development Organization (KEDO)
http://www.kedo.org/

The Korean Peninsula Energy Development Organization is a multinational organization established in 1994 from a nuclear disarmament agreement between the United States and the Democratic People's Republic of Korea. It works to implement energy-related projects in North Korea and provides support for international nuclear nonproliferation. The KEDO Web site provides updates on recent nuclear events associated with the Korean Peninsula, links to agreements and protocols, and information on nuclear safety.

Latin American Energy Organization (OLADE)
http://www.olade.org.ec

Formed in 1973, the Latin American Energy Organization is a multinational organization that offers political and technical support on energy issues to twenty-six countries in Latin America and the Caribbean. It seeks to promote energy security within the region while developing strategies for energy diversity and sustainability. The OLADE Web site has information on events and training for professionals in the energy industry, a National Energy Information System, which provides members access to energy statistics, and a list of publications produced by the organization.

Mine Safety and Health Administration (MSHA)
http://www.msha.gov/

The Mine Safety and Health Administration is a branch of the U.S. Department of Labor that is charged with developing and enforcing safety regulations and standards for the mining industry in the United States. Established in 1977 by the Federal Mine Safety and Health Act, MSHA seeks to reduce mining accidents, eliminate fatalities, minimize the health risks, and improve the working conditions for miners. The MSHA Web site provides up-to-date information on mine regulations, information on rights for professional miners, a reporting system for safety violations, and updates on important news stories. It also offers general education and training opportunities and a mine library.

National Association of State Energy Officials (NASEO)
http://www.naseo.org/

The National Association of State Energy Officials is a nonprofit organization that represents the interests of governor-appointed energy officials from each state and territory in the United States. The organization serves as an information source and collaborative forum for energy policies, programs, and issues in each state. The NASEO Web site contains information about state energy programs as well as contact information for energy offices in U.S. states and territories.

National Coal Council (NCC)
http://www.nationalcoalcouncil.org/

The National Coal Council serves as a policy advising body representing the coal industry to the U.S. Department of Energy. It provides advice on a number of issues, including policy reviews, scientific and engineering aspects, and opinions regarding research and development of coal technologies. The NCC Web site provides general and technical reports on the coal industry in the United States.

National Mining Association (NMA)
http://www.nma.org/

The National Mining Association is a trade organization for the U.S. mining industry. Formed in 1995, it provides a forum for the coal and mineral mining industries to promote their interests in the public policy process. The NMA Web site provides information on regulatory and legal issues, a compilation of statistics on the mining industry, and technology updates.

National Petroleum Council (NPC)
http://www.npc.org/

An advisory committee established in 1946 by a mandate from the U.S. government, the National Petroleum Council serves as an advisory council to the Secretary of Energy to represent the views of the oil and natural gas industry in energy matters. The NPC Web site provides links to yearly and special reports, downloadable presentations about global oil and gas resources, and information on NPC members.

National Renewable Energy Laboratory (NREL)
http://www.nrel.gov/

The National Renewable Energy Laboratory is a U.S. Department of Energy research facility that examines renewable energy technologies and their potential for meeting U.S. energy needs. The NREL serves as the primary research and development facility for renewable energy in the United States. The NREL Web site provides an overview of renewable energy and offers information on U.S. initiatives for renewable research and development and technology dissemination.

Natural Gas Supply Association (NGSA)
http://www.ngsa.org/

The Natural Gas Supply Association is a trade organization for the producers and marketers of natural gas in the United States. It promotes the natural gas industry in the areas of policy and business to enhance competitive markets for the production and transmission of domestic natural gas resources. The NGSA Web site provides information on social, policy, and regulatory issues important to the natural gas industry, offers facts and statistics, and contains updates on NGSA lobbying and professional activities.

Natural Resources Defense Council (NRDC)
http://www.nrdc.org/

The National Resources Defense Council is a nonprofit organization devoted to environmental protection. It works through scientific and legal channels to influence policy, promote an ethic of sustainability, and foster environmental justice. The NRDC Web site provides information on environmental issues, contains links to current environmental news stories, and provides suggestions for action at the individual and community levels.

New Buildings Institute (NBI)
http://www.newbuildings.org/

The New Buildings Institute is a nonprofit organization that works to promote the use of energy efficient measures in the construction of commercial buildings. It works collaboratively with utility groups to incorporate energy conservation guidelines in the design and construction of new buildings. The NBI Web site contains guidelines for lighting, heating, and cooling systems as well as building design codes.

New York State Energy Research and Development Authority (NYSERDA)
http://www.nyserda.org/

The New York State Energy Research and Development Authority is a public benefit corporation created to administer the Energy Smart Program in the State of New York. It provides energy-efficient services and implements research and development projects in the area of energy efficiency and environmental protection. The NYSERDA Web site provides information on incentives for creating energy efficient businesses and homes, descriptions of authority programs, and links to funding opportunities, state energy regulations, and news and data on energy prices and weather.

Northeast Energy Efficiency Partnerships (NEEP)
http://www.neep.org/

The Northeast Energy Efficiency Partnerships is a nonprofit organization that works to promote the use of energy-efficient products and services in the Northeast region of the United States. Formed in 1996, the NEEP has worked as an advocacy organization to increase the commercial availability of Energy Star products, promote information and technology exchange, and improve the efficiency of lighting, HVAC systems, and motors in industrial, commercial, and residential settings. The NEEP Web site provides information on efficiency standards for schools and businesses, offers materials for Building Operator Certification (BOC), and explains regional initiatives in energy efficiency.

Nuclear Energy Institute (NEI)
http://www.nei.org/

The Nuclear Energy Institute is a policy organization that promotes the beneficial use of nuclear energy throughout the world. Representing members in all aspects of the nuclear energy sector, from medicine to electricity generation, it provides a voice for the nuclear technologies industry, offers technical and business information, and provides up-to-date information about technology and policy. The NEI Web site contains nuclear statistics, an overview of nuclear technologies, and offers a science club that serves as an educational forum for children to learn about nuclear energy.

Nuclear Regulatory Commission (NRC)
http://www.nrc.gov/

The Nuclear Regulatory Commission is an independent agency in the federal government that oversees the civilian use of nuclear materials. It is charged with developing and enforcing regulations, licensing nuclear reactors, and ensuring radiation protection in the U.S. nuclear industry. It conducts inspections of nuclear facilities, performs research in the area of nuclear materials management, and oversees decommissioning operations. The NRC Web site provides information on nuclear reactors, nuclear materials, radioactive waste, and a searchable database of information on all nuclear facilities in the United States.

Office of Civilian Radioactive Waste Management (OCRWM)
http://www.ocrwm.doe.gov/

The Office of Civilian Radioactive Waste Management is a U.S. Department of Energy program charged with the task of managing a system for the long-term storage of spent fuel rods from nuclear reactors and high-level radioactive waste. Its main functions include developing a waste acceptance, storage, and transportation system and overseeing the construction and development of the Yucca Mountain Project, the site being considered as a permanent waste repository for high-level radioactive materials. The OCRWM Web site provides an information library, an overview of the issues involved in transporting and receiving radioactive wastes, and detailed information about the Yucca Mountain Project.

Office of Surface Mining (OSM)
http://www.osmre.gov/

The Office of Surface Mining (the full name of the agency is the Office of Surface Mining Reclamation and Enforcement, but they often shorten it to OSM) is a federal agency within the U.S. Department of the Interior responsible for overseeing the reclamation of land from mining activities. It is charged with developing standards of reclamation operation, providing financial aid to states and territories, and ensuring that land is restored for beneficial use subsequent to mining. The office seeks to harmonize domestic coal production with the values of environmental protection. The OSM Web site offers information about the regulation of active and abandoned mines, presents overviews of environmental research and technology associated with coal resources, and provides a reference center that includes statistics, laws and regulations, and publications.

Oil Depletion Analysis Center (ODAC)
http://www.odac-info.org/

The Oil Depletion Analysis Center is a UK-based organization that seeks to raise awareness about the issue of oil depletion. Formed in 2001, it provides information and educational materials to policymakers and the public about the geopolitical and economic consequences of a depleting oil supply. The ODAC Web site provides access to downloadable articles discussing the issue of peak oil and proposed solutions. The site also provides links to relevant Web sites with current news and information.

Organization for Petroleum Exporting Countries (OPEC)
http://www.opec.org/

The Organization for Petroleum Exporting Countries is an intergovernmental organization that represents the interests of the world's primary petroleum-producing countries. It works to coordinate production quotas and policies in member countries in an effort to stabilize global oil prices for producers and ensure a regular supply of petroleum to consuming countries. The OPEC Web site provides information on each of the eleven member countries, provides market updates for the petroleum industry, and contains links to news, publications, and OPEC seminars.

Petroleum and Natural Gas International Standardization (PNGIS)
http://www.pngis.net

Petroleum and Natural Gas International Standardization is an international committee created to develop international standards used in the petroleum and natural gas industries. It provides a database of regulations and standards used in these industries. The PNGIS Web site offers links to standards and regulations and lists of equipment manufacturers and service companies in the industries.

Petroleum Association of Japan (PAJ)
http://www.paj.gr.jp/index_e.html

The Petroleum Association of Japan is the trade association representing the refining and oil marketing industry in Japan. It works to promote industry interests to the government and provides information about oil supply, prices, forecasts, tax structures, standards of operation, and environmental protection. The PAJ Web site provides information about Japan's oil refining and marketing industry and recent oil statistics.

Petroleum Foundation of America (PFA)
http://www.ciglobal.com/pfa

The Petroleum Foundation of America is a charitable organization that seeks to offer incentives to the energy industry for providing services to economically disadvantaged groups of people. The PFA Web site provides information about the various service programs the group has initiated, including the Orphan Well Project and Gas for the Poor.

Petroleum Research Atlantic Canada (PRAC)
http://www.pr-ac.ca/

Formed in 2002, Petroleum Research Atlantic Canada is a not-for-profit organization dedicated to building and promoting petroleum research and development capacity for the Atlantic region of Canada. The PRAC Web site provides information on grant opportunities and downloadable reports, press releases, and presentations.

Petroleum Technology Alliance of Canada (PTAC)
http://www.ptac.org/

The Petroleum Technology Alliance of Canada is a not-for-profit organization that promotes research and development of Canada's oil and gas industry. It seeks to facilitate collaboration within the industry in research and development initiatives. The PTAC Web site provides resources and information in technical areas, announcements of workshops and other events, and links to industry, government, and research and development resources.

Petroleum Technology Transfer Council (PTTC)
http://www.pttc.org

Established in 1994, the Petroleum Technology Transfer Council is a not-for-profit organization that seeks to benefit independent producers of oil and natural gas. Its programs benefit the entire industry through the provision of technical reports and workshops to promote technology transfer within the industry. The PTTC Web site provides up-to-date technology summaries, downloadable reports for each region of the United States, and articles from newsletters and technology bulletins.

Petrotechnical Open Standards Consortium (POSC)
http://www.posc.org/

The Petrotechnical Open Standards Consortium, also known as the Petrotechnical Open Software Corporation, is a not-for-profit organization that promotes the sharing of exploration and production (E&P) information within the oil and natural gas industry. It offers information modeling and management specifications and has developed a data model, Epicentre, for the management of industry information. The POSC Web site provides overviews and updates for industry software products.

Pew Center on Global Climate Change
http://www.pewclimate.org/

The Pew Center on Global Climate Change seeks to provide a clear approach to the complex issue of climate change by engaging members in the business, scientific, and policymaking communities. It promotes collaboration and the use of sound science to analyze problems, inform policymakers, create business strategies, and educate all audiences about the issue of global climate change. The center's Web site offers in-depth background information about global warming, describes current business and policy initiatives, and provides an overview of key issues associated with climate change.

Production Engineering Association
http://www.peajip.org/

Members of the Production Engineering Association are upstream industry operators (that is, operators who deal with exploration, production, and processing and are "upstream" of the consumer) concerned with the promotion of hydrocarbons-production technology. It seeks to improve access and use of production technology on a global scale. The association's Web site is largely a resource for its members, which consist of suppliers to the industries as well as producers of petroleum and natural gas. It provides a list of members and links to those companies as well as information on forums and programs.

Radiation Effects Research Association (RERF)
http://www.rerf.or.jp/

The Radiation Effects Research Association is a research organization collaboratively managed by Japan and the United States. It was established to examine the effects of nuclear radiation on the survivors from the bombings of Hiroshima and Nagasaki. The RERF Web site provides overviews of association research programs, research results publications, data and archives, and links to Web sites that examine the effects of the bombings.

Regulatory Assistance Project (RAP)
http://www.raponline.org/

Formed in 1992, the Regulatory Assistance Project is a nonprofit organization that provides educational materials and research to public officials in the area of electric utility regulation. RAP offers workshops, newsletters, and information on issues such as renewable resources, electric utility restructuring, market development, green pricing, and demand-side management. It operates in forty-five states and a number of countries worldwide. The RAP Web site provides an overview of programs and offers numerous downloadable publications.

Resources for the Future (RFF)
http://www.rff.org/

Founded in 1952, Resources for the Future is a nonprofit research organization that provides economic and social analyses of environmental, economic, and energy issues. It applies the principles of economics in these research areas for the purpose of developing policies for the use and conservation of natural resources. It is a nonpartisan organization that shares its work with a number of government agencies, businesses, and organizations. The RFF Web site provides downloadable versions of the organization's reports on a variety of energy and other issues.

Rocky Mountain Institute (RMI)
http://www.rmi.org/

Established in 1982, the Rocky Mountain Institute is a nonprofit organization that provides information analysis on energy natural resource policy. It provides research and consulting services to businesses and communities operating on the local, national, and international levels. The RMI Web site offers information in a number of issue areas, including buildings and land, climate, transportation, water, and energy. It also offers an online discussion forum and a children's page.

Society of Petroleum Engineers
http://www.spe.org/

The Society of Petroleum Engineers is a professional organization for people with careers in exploration, development, and production of oil and natural gas resources. It provides information resources on a number of petroleum industry topics and serves as a forum for career development. The Web site offers information resources, handbooks, technical materials, and announcements for career advancement opportunities.

Society of Petrophysicists and Well Log Analysts (SPWLA)
http://www.spwla.org/

The Society of Petrophysicists and Well Log Analysts is a nonprofit corporation dedicated to advancing well logging and evaluation techniques used for oil, gas, and mineral exploitation. Formed in 1959, it promotes information dissemination and education for the petrophysical scientific community. The society's Web site provides chapter news, offers downloadable information and publications, and presents updates on conferences and events.

Solar Energy Industries Association (SEIA)
http://www.seia.org/

The Solar Energy Industries Association is a trade organization for the solar energy industry in the United States. It seeks to promote the expansion of solar technologies in global energy markets. The SEIA Web site provides up-to-date news about the solar energy industry, information about federal tax incentives for renewable energies, and publications about solar energy.

Solar Energy International (SEI)
http://www.solarenergy.org/

Solar Energy International provides technical and educational assistance to grassroots and development organizations that are interested in implementing sustainable energy programs. It offers workshops and training for the installation and use of solar and other renewable technologies. The SEI Web site contains information on the training programs offered and general information about renewable energy.

Southern Alliance for Clean Energy (SACE)
http://www.cleanenergy.org/

The Southern Alliance for Clean Energy is a nonprofit organization that seeks to advance the use of clean energy technologies in the southeastern United States. It works with state and local governments, community groups, utilities, and businesses to promote clean air policies and programs, expand the use of green power, enhance energy efficiency, and promote clean energy technologies. The SACE Web site provides extensive information on state programs for clean air, global warming, green power, and energy efficiency and offers suggestions for citizen involvement.

Stockholm Environment Institute (SEI)
http://www.sei.se/

The Stockholm Environment Institute is an independent research institute that examines issues of sustainable development. By working at the local, national, and international levels, the SEI seeks to build bridges between science and policy to develop strategies for sustainable initiatives. The SEI Web site offers an overview of institute programs, specific information on projects worldwide, and access to publications from SEI members.

Tellus Institute
http://www.tellus.org/

The Tellus Institute was formed in 1976 as a research organization specializing in the assessment of environment and development issues. Working at local, national, regional, and global levels, the institute provides program and policy analysis that links environmental, economic, and social aspects of development. Its analyses focus on energy, water, climate change, globalization, and sustainable development. The institute's Web site provides an overview of organizational strategies and initiatives and a searchable publications database.

United Nations Environment Programme (UNEP)
http://www.unep.org/

The United Nations Environment Programme is the environmental department of the United Nations. Established in 1972, it enables nations to develop sustainable communities without compromising future generations. It provides leadership to the global environmental movement by assessing environmental conditions at the global, regional, and national levels, providing support to environmental institutions, and promoting the transfer of information and technology. The UNEP Web site provides information resources for governments, scientists, journalists, businesses, and children. It also contains links to regional offices, recent news and events, and publications.

United Nations Framework Convention on Climate Change (UNFCCC)
http://unfccc.int

The United Nations Framework Convention on Climate Change is an international treaty that was ratified by 189 countries to address the issue of human-induced global climate change. The convention provides an overall framework for intergovernmental efforts to collect information on greenhouse gas (GHG) emissions and polices, adopt national strategies for GHG emission mitigation, and cooperate in international efforts to address the problem of global climate change. The UNFCCC Web site provides extensive background on the convention, GHG emissions data, national reports, the Kyoto Protocol, and the methods and science used in studying the issue of climate change.

U.S. Department of Energy (DOE)
http://www.energy.gov/

The Department of Energy is the primary federal energy agency in the United States. It is responsible for implementing U.S. energy policies, enhancing U.S. energy and nuclear security, promoting energy technologies, and managing the waste created from the nation's weapons programs. The DOE Web site provides a wide range of information on U.S. energy programs, goals, and activities.

U.S. Environmental Protection Agency (EPA)
http://www.epa.gov/

The primary mission of the U.S. Environmental Protection Agency is to protect the environment and human health in all areas of the environment (e.g., air, water, waste, etc.). The EPA's duties include developing and enforcing regulations, conducting environmental research, promoting environmental education, engaging in voluntary partnerships and programs, offering financial and technical assistance to states and territories, and publishing reports on its research and activities. The EPA Web site provides access to information about environmental protection in the United States, links to partnership and business opportunities, educational resources, links to environmental laws and regulations, and a children's information page.

World Coal Institute (WCI)
http://www.worldcoal.org/

The World Coal Institute is a nonprofit nongovernmental organization that seeks to advance the interests of the global coal industry. It provides a lobbying and information service for coal associations, organizations, and industries in the area of international energy and environmental policy. The WCI Web site provides general information about coal, statistics on coal use worldwide, and data on coal markets and pricing.

World Council for Renewable Energy (WCRE)
http://www.wcre.de/en

Founded in 2001, the World Council for Renewable Energy is an independent nongovernmental organization that seeks to enhance global discourse on renewable energy and promote renewable energy policies from the multinational to the community and individual levels. The WCRE Web site provides publications and press releases on efforts to expand the reach of renewable technology.

World Energy Council (WEC)
http://www.worldenergy.org

Established in 1924, the World Energy Council is a multinational, nongovernmental organization that provides information about energy resources and use. It collects and summarizes a vast amount of energy data from numerous countries, sponsors workshops and seminars on energy topics, and collaborates with energy organizations worldwide for the purpose of promoting sustainable supplies and uses of energy. The WEC Web site provides energy statistics and several downloadable energy publications that discuss up-to-date issues in both developing and industrialized countries.

World Energy Efficiency Association (WEEA)
http://www.weea.org/

Established in 1993, the World Energy Efficiency Association provides assistance to developing countries in the areas of energy technology, efficiency measures, and energy information services. It seeks to promote the diffusion of energy-efficiency efforts and coordinate the cooperation of international energy efforts. The WEA Web site contains a directory of energy service companies and organizations.

World Nuclear Association (WNA)
http://www.world-nuclear.org/

The World Nuclear Association works on a global scale to promote the development of nuclear power as a sustainable energy source for the future. It provides a forum for the distribution of technical and policy information. The WNA Web site provides data and information on all areas of the nuclear fuel cycle, access to international policy documents, an A-Z list of nuclear organizations around the world, and current information on nuclear issues.

World Resources Institute (WRI)
http://www.wri.org/

The World Resources Institute is a think tank that focuses on practical ways to protect the Earth's environment. It seeks to reverse ecosystem damage, increase democratic participation in environmental decisions, improve environmental stewardship in the private sector, and mitigate the adverse effects of climate change.

The WRI Web site contains information on a variety of environmental issues, offers reports aimed at individuals in the government and business communities, and provides a forum for ideas on how to mitigate environmental damage in all aspects of human life, from capital markets to natural ecosystems.

8

Resources

Introduction

This chapter provides an overview of the numerous print and nonprint resources on the topic of energy. It is important to note that this is a selective list of resources. The topic of energy use worldwide is expansive; hence, it would be impossible to create an exhaustive list of information resources. This chapter is divided into three subject areas. The first, general energy, offers a bibliography of references about energy concepts, energy history, and energy dynamics in different world regions. Next are references for specific energy resource categories, such as oil and gas, nuclear energy, and renewable energy. The third category provides references in the area of social and environmental problems from energy use.

General Energy

Books

Bailer, U. 1999. *Oil and the Arab-Israeli Conflict, 1948–63.* New York: St. Martin's Press. 282 pp.

This book discusses the development of Israel's energy resources during the first fifteen years of statehood. It provides an overview of the political and economic struggle for Israel to develop oil resources and infrastructure in the context of the Arab-Israeli conflict. It offers background on the politics of the Middle East and is a good reference for students who need to understand the importance of oil resources to politics and development in the region.

Bent, R., L. Orr, and R. Baker, eds. *Energy: Science, Policy, and the Pursuit of Sustainability.* **Washington, DC: Island Press. 257 pp.**

This volume provides an overview of general aspects of energy, describes the environmental and economic problems associated with current levels of energy use, discusses energy policy and economics, and puts sustainable growth in a context of energy use. This is a useful reference for students who wish to gain a broad concept of energy and society.

Davis, H. D. 2001. "Energy on Federal Lands." In *Western Public Lands and Environmental Politics,* **2nd ed., edited by C. Davis, 141–168. Boulder, CO: Westview Press.**

This chapter provides a history of energy policy on federal lands since 1975. It provides overviews of resource extraction, energy production, and environmental and land management policies enacted under the Nixon, Ford, Carter, Reagan, H. W. Bush, and Clinton administrations. This is a good reference for anyone interested in understanding the impact of federal policies on western U.S. public lands.

Dienes, L., and T. Shabad. 1979. *The Soviet Energy System: Resource Use and Policies.* **Washington, DC: V. H. Winston & Sons. 298 pp.**

This book provides an overview of energy use and development in the Soviet Union during the first three-quarters of the twentieth century. It includes information about fossil fuel, hydropower, and nuclear energy resources in the Soviet Union. It is particularly useful for understanding how the Soviet Union developed its energy resources and rose to be a powerful nation in the world.

Fanchi, J. R. 2005. *Energy in the 21st Century.* **Hackensack, NJ: World Scientific. 243 pp.**

This book, written in nontechnical terms, provides an overview of fossil and nonfossil energy technologies and their projected roles.

It examines the energy options of fossil fuels, nuclear power, solar energy, wind, water, biomass, and hydrogen. It also discusses electricity generation, economics, the environment, and energy forecasts. This book is a useful reference for those who would like to understand the future direction of various energy resources.

Hodgson, P. E. 1999. *Nuclear Power, Energy and the Environment.* **London, England: Imperial College Press. 205 pp.**

This book discusses energy use within the context of energy crises and future energy resources. It provides an overview of both renewable and nonrenewable resources, nuclear power and the operation of nuclear reactors, the decision-making elements involved in choosing the best energy source, environmental effects of energy use, and the political and moral aspects of energy policies. This book is a good reference for students who want a broad overview of energy and its impacts to society.

Hoffman, G. W. 1985. *The European Energy Challenge: East and West.* **Durham, NC: Duke University Press. 207 pp.**

This book provides an overview of energy policies that have been implemented across Eastern and Western Europe since the energy crises of the 1960s and 1970s. It analyzes the energy dilemma throughout Europe, examines trends of energy diversification and efficiency, and discusses the role of the former Soviet Union as an increasingly important supplier of energy resources. This book is a good resource for students and scholars who wish to understand the political and economic dynamics of late twentieth-century European energy policy.

Howes, R., and A. Fainberg, eds. 1991. *The Energy Sourcebook: A Guide to Technology, Resources, and Policy.* **New York: American Institute of Physics. 536 pp.**

This book provides an overview of the state of energy resources within the context of energy crises. It explains the technological development and exploitation of fossil fuels, nuclear, solar, hydroelectric, geothermal, ocean, biomass, and wind energy resources. It also describes how these resources are converted into electricity as well as used in the sectors of agriculture, transportation, and commercial and residential buildings. This book is useful for students who require a general technical overview of how resources are converted into useful energy services in society.

Kapstein, E. B. 1990. *The Insecure Alliance: Energy Crises and Western Politics since 1944.* Oxford, England: Oxford University Press. 257 pp.

This book provides a historical overview of energy dynamics since World War II. It outlines geopolitical alliances and describes the dynamics of several important energy events including Europe's postwar reconstruction, the impact of the Marshall Plan, the Suez Crisis, the establishment of the Organization of Oil Producing Countries (OPEC), the Organization for Economic Cooperation and Development (OECD) and the International Energy Agency (IEA), the oil embargo of 1973–1974, and the Iran-Iraq War. This is a good book for students who wish to understand how post–World War II events have shaped global energy dynamics and relations.

Laitos, J. G., and J. P. Tomain. 1992. *Energy and Natural Resources Law in a Nutshell.* St. Paul, MN: West Publishing. 554 pp.

This book provides a quick, ready reference for natural resource and environmental law. It discusses economics, natural resource extraction, and water, timber, and environmental laws, and links these discussions to energy resource extraction, production, and transmission regulations. This is a useful reference for students and scholars who need to understand legislative, regulatory, and case law precedent for energy dynamics in U.S. society.

MacKerron, G., and P. Pearson, eds. 2000. *The International Energy Experience: Markets, Regulations, and the Environment.* London, England: Imperial College Press. 375 pp.

This book examines the dynamics of energy markets in the context of market liberalization and environmental concerns. It provides an overview of world energy and oil markets and examines national energy structures, regulatory aspects of oil and natural gas utilities, how the goals of economic efficiency and environmental stewardship can be achieved simultaneously, and the role of renewable energy. This book is a good reference for students who want to understand the dynamics and future challenges of global energy markets.

Manning, R. A. 2000. *The Asian Energy Factor: Myths and Dilemmas of Energy, Security and the Pacific Future.* **New York: Palgrave. 246 pp.**

This book provides an overview of energy dynamics in Asia, with chapters devoted to China, India, the Korean Peninsula, Japan, and the countries in Southeast Asia. It offers both economic and political perspectives on the highest-populated and fastest-growing region in the world. This book is useful for students who wish to understand current and projected energy use in Asia and its potential global consequences.

Melosi, M. V. 1985. *Coping with Abundance: Energy and Environment in Industrial America.* **Philadelphia, PA: Temple University Press. 355 pp.**

This book is an overview of U.S. energy history from the industrial revolution to the 1980s. It provides a detailed account of energy transitions from wood to coal to oil and discusses the evolution of electric utilities, the impacts of American consumerism and the automobile, the role of U.S energy sources in World Wars I and II, and the impact of the energy crisis in the 1970s. This is an excellent reference for students who want to understand energy dynamics in the United States since 1820.

Millán, J., and N.-H. M. von der Fehr, eds. 2003. *Keeping the Lights On: Power Sector Reform in Latin America.* **Washington, DC: Inter-Development Bank. 383 pp.**

This book is an analysis of electricity market reforms that occurred throughout Latin America in the 1990s. It provides case studies of utility reforms in Colombia, Honduras, and Guatemala and examines the political, social, and economic aspects of electricity markets throughout Latin America. This is a good reference for upper level students who want to understand the issues of market reform and electricity provision in developing Latin American countries.

Miller, E. W., and R. M. Miller. 1993. *Energy and American Society: A Reference Handbook.* **Santa Barbara, CA: ABC-CLIO. 418 pp.**

This book serves as a general reference guide to energy use in the United States. It provides an overview of fossil and nonfossil energy resources, a history of U.S. energy use, a chronology of important energy events, facts and figures, a list of organizations, and energy resources available. It is a useful reference for students wanting to understand general aspects of U.S. energy use.

Mitchell, J. V., P. Beck, and M. Grubb. 1996. *The New Geopolitics of Energy*. London, England: The Royal Institute of International Affairs, Energy and Environmental Programme. 196 pp.

This book provides an overview of global energy dynamics in the late 1980s and early 1990s. It links politics and economics to energy demand on a regional and global scale and includes overviews of the nuclear issue and climate change. It serves as a useful guide for understanding energy dynamics in the Middle East, Russia, East Asia, and the non-OPEC world. It is a useful reference for students wishing to gain a multidisciplinary perspective on global energy use.

Paik, K. W. 1995. *Gas and Oil in Northeast Asia: Policies, Projects, and Prospects*. London, England: Royal Institute of International Affairs, Energy and Environmental Programme. 274 pp.

This book provides an overview of the oil and gas industry in Northeast Asia. It examines the political and economic aspects of energy development in Russia, China, Japan, South Korea, and the gas and oil fields of Sakhalin, Sakha, and Tarim. This book is a good reference for students who want to understand the energy dynamics in Northeast Asia and their importance to the region's development.

Ramage, J. 1997. *Energy: A Guidebook*. Rev. ed. Oxford, England: Oxford University Press. 394 pp.

This book covers fundamental aspects of energy production, conversion, distribution, and consumption in society. It provides overviews of renewable and nonrenewable energy resources, details the basic concepts behind how these resources are extracted and converted to useful energy by society, and discusses and summarizes the environmental impacts of energy use. This is a useful resource for anyone interested in understanding general energy concepts.

Rose, D. J. 1986. *Learning about Energy.* **New York: Plenum Press. 506 pp.**

This book offers a fundamental overview of energy use in society. It discusses economic and environmental considerations of energy use, the importance of conservation, fossil and nonfossil resources used for energy provision in society, and the electricity and energy storage systems that have been developed. This book is a useful reference for students who want an easily understandable reference about the technical aspects of energy in society.

Rosenbaum, W. A. 1993. "Energy Policy in the West." In *Environmental Politics and Policy in the West,* **edited by Z. A. Smith, 177–199. Dubuque, IA: Kendall/Hunt Publishing.**

This chapter discusses energy issues impacting the western United States. It provides an introduction to oil and gas drilling, renewable energy potentials, relations between state and federal governments in the context of energy resources, an overview of the nuclear waste storage controversy, and environmental issues. This chapter is useful for students who need a succinct overview of energy issues in the United States.

Siddayao, C. M. 1986. *Energy Demand and Economic Growth: Measurement and Conceptual Issues in Policy Analysis.* **Boulder, CO: Westview Press. 127 pp.**

This book describes the link between energy, economics, and development. It explains energy measurement and analysis, describes how energy demand is calculated, details the factors that contribute to energy use in processes of development, and explains the utility of different energy indicators, such as energy intensity. It is a good summary reference for policymakers, scholars, and students who wish to understand the fundamental, macroeconomic concepts in energy demand.

Smil, V. 1988. *Energy in China's Modernization: Advances and Limitations.* **Armonk, NY: M. E. Sharpe. 250 pp.**

This book discusses China's potential energy resources, extraction and utilization of energy in rural and urban areas, goals and strategies outlined for energy distribution and use in industrialization efforts, and potential environmental considerations that need to be made during this transition. It provides a useful resource for students and scholars who wish to understand energy resource distribution (both renewable and nonrenewable), the economic dynamics of rural and urban growth, and overall energy dynamics in China.

Smil, V. 1994. *Energy in World History.* **Boulder, CO: Westview Press. 299 pp.**

This book offers a historical account of energy use throughout human history. It discusses the development of energy technologies in all aspects of society, most notably in the areas of agriculture, transportation, and grain milling. The majority of the book is devoted to preindustrial energy resources and technology, with only a chapter discussing fossil energies, making this an excellent guide for understanding energy dynamics prior to the Industrial Revolution.

Smil, V. 1999. *Energies: An Illustrated Guide to the Biosphere and Civilization:* **Cambridge, MA: MIT Press. 210 pp.**

This book demonstrates the importance of energy in all aspects of life. It discusses energy flows found in the Sun and the Earth, plants and animals, food and metabolism, and in pre- and postindustrial societies. This book is a useful reference for students and scholars who want a broad picture of how energy flows throughout all life systems.

Smil, V. 2003. *Energy at the Crossroads: Global Perspectives and Uncertainties.* **Cambridge, MA: MIT Press. 427 pp.**

This book provides a comprehensive overview of energy issues and examines their implications for future energy dynamics. It examines society's energy use in all sectors; describes links between energy and economics, the environment, and war; examines the potential future of fossil and nonfossil fuels; and focuses on the importance of energy efficiency in the future. This book is a thorough investigation into the complexity of energy dynamics and offers a critical view of the potential solutions to energy problems.

Stares, P. B., ed. 2000. *Rethinking Energy Security in East Asia.* **Tokyo: Japan Center for International Exchange. 207 pp.**

This book examines energy security scenarios in East Asia in light of the region's growing energy demands. Energy experts from seven Asian countries analyze energy issues, such as an expansion of nuclear energy and a reliance on oil and gas imports, in China, Japan, South Korea, Taiwan, the Association of Southeast Asian Nations, and Russia. The role of the United States in Asian energy policy is also discussed. This book is a good reference for students who wish to understand the increasingly important role of Asia in global energy dynamics.

Stoker, H. S., S. L. Seager, and R. L. Capener. 1975. *Energy: From Source to Use.* **Glenview, IL: Scott Foresman and Company. 337 pp.**

This book provides an overview of energy within the context of energy crises. It discusses the dynamics of fossil, nuclear, and renewable energy resources, examines the 1973 energy crisis within the context of these resources, and explains how energy conservation can provide a viable solution to the problems associated with energy provision and use. This book serves as a general resource for those who want to understand basic energy technology, the nature of energy crises, and the option of energy conservation.

Teixeira, M. G. 1996. *Energy Policy in Latin America: Social and Environmental Dimensions of Hydropower in Amazonia.* **Aldershot, Hants, England: Avebury. 348 pp.**

This book provides an overview of hydropower development in the Amazon region of South America. It describes the environmental and social impacts of large dams and the long-term costs to the Amazon River ecosystems and its people. This is a good reference for students who need to understand the impact of hydroelectricity in a region where water resources supply a significant amount of energy.

Periodicals, Journals, and Newsletters

Africa Energy Intelligence
CMS Business Information

2a Altons House Office Park, Gatehouse Way,
Aylesbury, HP103XU, England
   ISSN: 1635-2742
   http://www.biz-lib.com/index.html

This biweekly newsletter provides business news to the African mining, oil, gas, and utility industries. It offers up-to-date information on company mergers, privatization, and energy projects across Africa.

Applied Energy
Elsevier Inc. (Branch office)
30 Corporate Drive, 4th floor
Burlington, MA 01803
   ISSN: 0306-2619
   http://www.elsevier.com/

This scholarly journal contains papers and reviews that report on research and development of energy conversion, conservation, and management. It is intended for energy engineers and researchers in the areas of conservation and alternatives energies.

Energy: The International Journal
Elsevier Inc. (Branch office)
30 Corporate Drive, 4th floor
Burlington, MA 01803
   ISSN: 0360-5442
   http://www.elsevier.com/

This scholarly journal offers an interdisciplinary focus on energy policy and program assessment and management. It is intended to be a resource for energy planners, researchers, and industrial producers and consumers.

Energy and Buildings
Elsevier Inc. (Branch office)
30 Corporate Drive, 4th floor
Burlington, MA 01803
   ISSN: 0378-7788
   http://www.elsevier.com/

This scholarly journal provides a forum for the presentation of research and practice in the area of energy-efficient building design, heating and cooling systems, and energy conservation. It is intended for energy planners, policymakers, architects, energy engineers, and the building community.

Energy Compass
Energy Intelligence
5 East 37th Street, 5th Floor
New York, NY 10016
    ISSN: 0962-9270
    http://www.energyintel.com/

This weekly publication offers news and analysis of geopolitical events important in the energy industry. It is intended for business professionals, government officials, and energy market analysts.

Energy and Electricity Forecast World
Economist Intelligence Unit (EIU)
111 West 57th Street
New York, NY 10019
    http://www.eiu.com/

This periodical provides information on energy and electricity industries, including descriptions of key players, supply and demand overviews, energy indicators, and five-year forecasts. It is a publication relevant to businesses and governments in assessing current trends in energy use and markets.

Energy Economics
Elsevier Inc. (Branch office)
30 Corporate Drive, 4th floor
Burlington, MA 01803
    ISSN: 0306-2619
    http://www.elsevier.com/

This scholarly journal publishes research in the areas of energy finance, economic theory, regulatory and computational economics, statistics, and modeling. It is directed toward energy economists, financial analysts, and academic researchers.

Energy Law Journal
Energy Bar Association
1020 19th St., N.W., Suite 525

Washington, D.C. 20036
    ISSN: 0270-9163
    http://www.eba-net.org

This biannual scholarly law journal provides articles in the field of energy law written by and for practitioners, academics, judges, and federal agency officials.

Energy Magazine
Business Communications Company
40 Washington St., Suite 110
Wellesley, MA 02481
    http://www.bccresearch.com/

This quarterly newsletter contains articles written by industry experts on a variety of energy topics, including energy exploration, economics, utilities, technology, renewable energy, and conservation.

Europe Energy
Europe Information Service
Avenue Adolphe Lacomblé, 66-68
B-1030 Brussels, Belgium
    ISSN: 0772-1528
    http://eisnet.eis.be/

This newsletter provides information and news about the energy sector in Europe, including policy, statistics, research, and technology.

Journal of the Institute of Energy
Institute of Energy
61 New Cavendish Street
London W1G 7AR, United Kingdom
    ISSN: 0144-2600
    http://www.energyinst.org.uk/

This quarterly scholarly journal is directed toward energy scientists and engineers. It provides up-to-date articles on advances in energy and fuel technologies.

Films and Videorecordings

Conservation of Energy: Potential Energy

The California Institute of Technology and the Corporation for Community College Television. Published by Intellimation, 1987. Annenberg/CPB Project produced by P. F. Buffa and directed by M. Rothschild.
VHS, 58 minutes

This film provides an overview of the physical laws of energy conservation and potential energy. It describes the concepts and mathematics that underlie the fundamental laws of energy in nature. This film is a good supplement to introductory physics.

Energy: Nature's Power Source
Advanced Video Productions, Inc.
Meridian Education Corporation, 1998.
VHS, 18 minutes

This film provides a brief overview of different energy sources, how each source is utilized and its positive and negative aspects, the projected reserves and supplies, and provides a succinct summary of energy use.

Energy and Society
Hawkhill Associates, 1990.
Written and directed by B. Stonebarger.
VHS, 35 minutes

This film provides an overview of the scientific and technological feats that have been accomplished in order for society to harness energy resources. Major themes in the film include the difference between high- and low-energy societies, the benefits of high-energy societies, scientific and technological aspects of energy, the negative impacts of burning fossil fuels, and the importance of increasing energy efficiency.

The Politics of Power
Frontline and the Center for Investigative Reporting. Published
by PBS Video, 1992.
Produced by J. Legnitto and correspondent N. Kotz.
VHS, 58 minutes

This film is an investigative report of the National Energy Policy
(NEP) passed in 1992 under George H. W. Bush. It examines the
role of special interests in the policy, the issue of energy security,
and offers a critical overview of the intention of the NEP.

Search for Common Ground on Energy
Common Ground—WNYC Production. Published by Common
Grounds Production, 1989.
Produced by W. B. Shanley and directed by J. Chiappardi.

This film examines the issue of energy crisis from a number of
different perspectives. It provides insight from energy producers
and consumers, environmental groups, economists, and lobbyists.

Voltage, Energy and Force, the Electric Battery

The California Institute of Technology and the Corporation for
Community College Television. Published by Intellimation, 1987.
Annenberg/CPB Project, produced by P. F. Buffa.
VHS, 58 minutes

This film provides an overview of the physical and mathematical
concepts of energy and force, the physics behind how an electric
battery operates, and how chemical energy is converted to electrical energy. It is a good supplement to introductory physics.

Databases and Internet Resources

Enerdata: Global Energy Intelligence
http://www.enerdata.fr/enerdatauk/index.html

Enerdata is an independent consulting and information services
company that offers a variety of standard and customized databases for different energy resources (fossil and nonfossil energy,

renewables, etc.), global and regional energy markets, energy efficiency measures, and key energy statistics.

Energy Citations Database
U.S. Department of Energy
http://www.osti.gov/energycitations/

Energy Citations database provides access to scientific and technical energy publications. It offers bibliographic records for energy-related abstracts in chemistry, physics, engineering, climatology, geology, and related fields.

EnergyFiles! Database
Energy Science and Technology Virtual Library
U.S. Department of Energy
http://www.osti.gov/energyfiles/

This Web site provides access to over 500 energy databases and Web sites that examine technical and scientific aspects of energy use. It offers access to energy-related science and technology information, accessible and efficient data retrieval mechanisms, and links that facilitate energy-related electronic research.

Energy Information Administration (EIA): Official Statistics from the U.S. Government
http://www.eia.doe.gov/

This Web site provides an abundant amount of energy and environmental information. It offers international and U.S. statistics on energy resource production, use, and consumption. It provides market data for major energy commodities; analysis of energy use in each sector; weekly, monthly, and annual reports about energy use; basic energy information; and historical energy data.

International Energy Agency (IEA)
Energy Information Centre
http://www.iea.org/Textbase/subjectqueries/index.asp

This Web site offers a searchable database of information on all aspects of energy use worldwide. The site provides access to information for all countries and eleven regional divisions. Topics include clean fossil fuels technologies, climate change, greenhouse gas (GHG) emissions and trading mechanisms, energy efficiency, market reports, sustainable development, energy policy, and renewable technology.

Platt's
McGraw-Hill Company
http://www.platts.com/

Platt's Web site provides access to in-depth, up-to-date analyses of global and regional markets for all energy resources and commodities, news on important energy events, energy statistics and reports, and updates regarding energy futures and financing.

World Energy Database
Energy Technology Data Exchange (ETDE)
http://www.etde.org/

World Energy Database provides information on energy technologies and research worldwide. It contains over 3.7 million abstracts and 175,000 full-text links to research and development publications on basic energy sciences, environmental issues of energy production and consumption, climate change, and renewable energy, nuclear, coal, and fossil fuel resources.

Energy Resources

Books

Berinstein, P. 2001. *Alternative Energy: Facts, Statistics, and Issues.* Westport, CT: Oryx Press. 208 pp.

This book provides an overview of alternative energy resources: solar, biomass, wind, ocean, fusion, geothermal, and hydrogen. It discusses overall energy issues and the economics of alternative energy sources. It examines issues of energy conservation and efficiency, energy storage systems, fuel cells, and the role of renewables in the transportation sector. This book is a useful resource for students who wish to understand renewable energy technology.

Borowitz, S. 1999. *Farewell Fossil Fuels: Reviewing America's Energy Policy.* **New York: Plenum. 220 pp.**

This book examines the technological aspects of resource extraction and exploitation of fossil and nonfossil energy resources. It covers oil, coal, natural gas, nuclear energy, solar, wind, biomass, geothermal, hydrogen fuel cells, and other types of renewable resources. This book is intended for those who wish to understand and influence the use of renewable energy technologies.

Boyle, G., ed. 1996. *Renewable Energy: Power for a Sustainable Future.* **Oxford, England: Oxford University Press. 479 pp.**

This book discusses renewable energy sources that are exploited by society. Each chapter is devoted to a resource, providing detailed information about its energy potentials, historical and current use, technological developments, economic feasibility, and case studies that illustrate particular areas where the resource is used. This is an excellent reference for students interested in expanding their knowledge about all types of renewable energy.

Brennan, T. J. 2002. *Alternating Currents: Electricity Markets and Public Policy.* **Washington, DC: Resources for the Future. 210 pp.**

This book provides an overview of the electricity industry in the United States and the impacts of electricity market restructuring. It examines the current market structure, reviews international experiences in restructuring, offers a detailed explanation of California's electricity market reforms, and examines a number of policy issues such as rate regulation, enhancing competition, the roles of state and federal governments, and how electricity can be reliable amid restructuring efforts. This book is an excellent reference for students who need to understand the dynamics of utility regulation and electricity market restructuring.

Cohen, B. L. 1990. *The Nuclear Energy Option: An Alternative for the 90s.* **New York: Plenum Press. 338 pp.**

This book presents an argument in support of nuclear power. It explains the various risks associated with nuclear power plants, radiation exposure, and radioactive waste; details what went wrong at Chernobyl; and addresses the issue of nuclear proliferation. The book aims to put the risks of nuclear power in perspective. This is a good reference for those who seek a pronuclear opinion.

Edinger, R., and Sanjay Kaul. 2000. *Renewable Resources for Electric Power: Prospects and Challenges.* **Westport, CT: Quorum. 154 pp.**

This book examines the feasibility of using renewable resources for the production of electric power. It describes the restructuring of the electricity industry, discusses the challenges for renewable energy in the context of market liberalization, and examines how solar, wind, and microhydroelectric technologies can support small-scale electricity generation. This book is a useful resource for understanding the benefits, risks, and feasibility of integrating renewable technologies into existing electricity infrastructure.

Freese, B. 2003. *Coal: A Human History.* **Cambridge, MA: Perseus Publishing. 308 pp.**

This book offers a historical overview of the extraction, production, and use of coal. It examines the development of coal in Great Britain, the rise of "King Coal," the pollution created from coal use, the social problems existing in the past and present in coal mines, and the importance of coal in today's society. This book is a useful reference for students who want to understand the historical evolution of coal and its impacts to society.

Gillespie, K., and Clement Moore Henry, eds. 1995. *Oil in the New World Order.* **Gainesville: University Press of Florida. 339 pp.**

This book discusses political and economic concerns associated with the supply and demand of oil. It examines global oil markets, energy dynamics in the Middle East and Russia, and offers case studies of important petroleum producing countries. This book is a good reference for students and scholars who wish to understand the contentious, political dynamics of global oil provision.

Gorman, H. S. 2001. *Redefining Efficiency: Pollution Concerns, Regulatory Mechanisms, and Technological Change in the U.S. Petroleum Industry.* **Akron, OH: University of Akron Press. 451 pp.**

This book provides an overview of the evolution of pollution control regulations in the petroleum industry. It discusses the historical record of pollution concerns in the extraction, production, and refining of petroleum; how industry leaders responded to these concerns; and the technical changes that have occurred in the industry throughout history. This book is a useful reference for those who wish to understand the history and dynamics of pollution regulation in the petroleum industry in the twentieth century.

Hatch, M. T. 1986. *Politics and Nuclear Power: Energy Policy in Western Europe.* **Lexington: University Press of Kentucky. 219 pp.**

This book examines energy policies and issues in Western Europe since World War II. In particular, it examines the development of nuclear energy resources in Germany, France, and the Netherlands and why this development has resulted in different energy outcomes in each country. This book is useful for students and scholars who wish to understand the dynamics of energy development in Western Europe in the latter half of the twentieth century.

Josephson, P. R. 2000. *Red Atom: Russia's Nuclear Program from Stalin to Today.* **New York: W. H. Freeman. 352 pp.**

This book provides a historical overview of nuclear development in post–World War II Russia and throughout the Cold War. It profiles the scientists and politicians involved in the Russian nuclear age, analyzes the Chernobyl disaster, and reviews issues of waste storage and abandoned nuclear stockpiles. This book is a good reference for students who want to understand the development of nuclear technology and issues that impact the utilization of atomic power in Russia.

Karekezi, S., and T. Ranja. 1997. *Renewable Energy Technologies in Africa.* **London, England: Zed Books. 269 pp.**

This book, published in association with the African Energy Policy Research Network and the Stockholm Environment Institute, is an overview of renewable energy initiatives in eastern and southern Africa. It examines the development of biomass, solar, wind, and hydropower resources and the sustainable use

of energy supplies. It also considers the political and socioeconomic conditions, financing and energy management schemes,
and equity, access, and environmental sustainability of Africa's
energy sector. This book is a good reference for students who
want to understand energy development and sustainability in
Africa.

Kellow, A. J. 1996. *Transforming Power: The Politics of Electricity Planning.* **Cambridge, England: Cambridge University
Press. 229 pp.**

This book analyzes how the electricity industry has responded to
energy crises and environmental concerns in Canada, Australia,
and New Zealand. It provides a comparative overview of how
each country has planned and developed their energy resources,
utility markets, and transmission systems for the electricity industry. This book is a good reference for students who wish to understand the development of electricity infrastructure and policy
in industrialized countries.

Kruschke, E. R., and B. M. Jackson. 1990. *Nuclear Energy Policy:
A Reference Handbook.* **Santa Barbara, CA: ABC-CLIO. 245 pp.**

This book serves as a general reference guide for understanding
nuclear energy. It examines the arguments in support of and
against developing nuclear power, provides detailed information
about the history and current state of the nuclear power industry,
and serves as a guide to documents, organizations, and resources
associated with the nuclear issue. It is a useful reference for
people who want to learn more about nuclear energy.

League of Women Voters Education Fund. 1993. *The Nuclear
Waste Primer.* **New York: N. Lyons Books. 170 pp.**

This book provides an overview of the issues associated with nuclear waste. It includes general information about nuclear waste,
the impacts of radiation, responsibilities of state and federal players, the distinction between high-level and low-level radioactive
waste, transportation and liability issues, and waste that is produced from defense programs. This is a good reference for students who wish to know more about how radioactive waste is
generated and managed in the United States.

Morris, R. C. 2000. *The Environmental Case for Nuclear Power: Economic, Medical, and Political Considerations.* St. Paul, MN: Paragon House. 192 pp.

This book examines the benefits of nuclear power. It reviews the problems associated with a reliance on fossil fuels, discusses the many benefits of nuclear technology, examines the problem of nuclear waste, and argues the need to enhance nuclear capabilities. It is a useful reference for students who wish to understand arguments in support of nuclear energy.

Sabin, P. 2005. *Crude Politics: The California Oil Market, 1900–1940.* Berkeley: University of California Press. 307 pp.

This book describes the development of the oil industry in California in the first half of the twentieth century. It highlights the role of politics and law in shaping oil markets, examines the relationship between business and government in promoting the growth of the industry, and presents a history of the development of the oil industry in the western United States. This book is a useful resource for students who wish to understand the historical development of the oil industry and the role that politics played in its market dominance.

Sampson, A. 1975. *The Seven Sisters: The Great Oil Companies and the World They Shaped.* New York: Viking Press. 334 pp.

This book offers a detailed overview of the rise of the oil industry around the world by focusing on the seven large oil companies that established the global market. It provides a readable glimpse into the complex world of energy markets, the closed door policies (e.g., the Red Line Agreement, the As-Is Agreement, etc.) established by oil companies, and how a selective number of companies came to dominate the global petroleum industry. This book is a good reference for those who wish to understand the evolution of oil markets since the late 1800s.

Scamehorn, H. L. 2002. *High Altitude Energy: A History of Fossil Fuels in Colorado.* Boulder: University Press of Colorado. 244 pp.

This book provides a historical overview of oil and natural gas production and coal mining in the state of Colorado through an examination of the social and economic history of energy extraction and production industries. This book is a good reference for students who need to understand the development of fossil energy industries in the western United States. Although the events in the book are specific to Colorado, the economic and social conditions described are similar to those in other Western mountain states.

Shojai, S., ed. 1995. *The New Global Oil Market: Understanding Energy Issues in the World Economy*. Westport, CT: Praeger. 263 pp.

This book examines dynamics in global oil markets. It provides an overview of oil reserves, patterns in production and consumption, oil pricing, futures in oil markets, and marketing. It describes the role of important players such as the Organization of Oil Producing Countries (OPEC) and the International Energy Agency (IEA), considers the socioeconomic impact of oil markets in producing and consuming countries, and looks at the environmental effects of current oil market structure. This book is a useful resource for students who wish to understand the fundamentals of global oil markets and current trends within the industry.

Wu, K. 1995. *Energy in Latin America: Production, Consumption, and Future Growth*. Westport, CT: Praeger. 310 pp.

This book provides an overview of energy growth and supply in Latin America. Chapters are divided by individual countries in Latin America, and each gives a description of energy supply, petroleum consumption, natural gas utilization, and policy issues impacting the oil and natural gas industries. This book is a good reference for those interested in oil and natural gas dynamics in Latin America.

Periodicals, Journals, and Newsletters

Coal Age
Mining Media Inc.
8751 East Hampden Ave, Ste B-1
Denver, CO 80231

ISSN: 1091-0646

This monthly journal is a resource for the coal mining industry. It provides up-to-date information on mining equipment and technology, methods and techniques, and safety issues.

International Journal of Coal Geology
Elsevier, Inc.
30 Corporate Drive, 4th floor
Burlington, MA 01803

>   ISSN: 0166-5162
>   http://www.elsevier.com

This academic journal publishes information on the science of coal geology. It provides up-to-date research in the field, surveys of coal resources, and literature and book reviews for an audience of geologists, analysts in the coal industry, and professionals studying coal science.

Oil and Gas Journal
PenWell Petroleum Group
1700 West Loop South, Suite 1000
Houston, TX 77027

>   ISSN: 0030-1388

This journal on the oil and gas industry provides market news, analysis of important events, updates on recent technology, and statistics on national and international markets.

Progress in Nuclear Energy
Elsevier, Inc.
30 Corporate Drive, 4th floor
Burlington, MA 01803
>   ISSN: 0149-1970
>   http://www.elsevier.com

This international academic review journal publishes articles on nuclear science, engineering, and aspects of the power industry, including safety, economics, and fuel management.

Renewable Energy
Elsevier, Inc.
30 Corporate Drive, 4th floor
Burlington, MA 01803

ISSN: 0960-1481
http://www.elsevier.com

This academic journal publishes original research papers, reviews, and reports on new developments in the field of renewable energy. Topics include solar, wind, biomass, wave and tide, and minihydropower technologies, energy conservation and efficiency, and socioeconomic issues related to energy management.

Solar Energy
Elsevier, Inc.
30 Corporate Drive, 4th floor
Burlington, MA 01803

ISSN: 0038-092X
http://www.elsevier.com

This is the official academic journal of the International Solar Energy Society. It publishes articles concerning all aspects of solar energy applications, research, development, measurement, and policy. It appeals to a wide audience of scientists, engineers, architects, and economists.

Solar Energy Materials and Solar Cells
Elsevier, Inc.
30 Corporate Drive, 4th floor
Burlington, MA 01803

ISSN: 0927-0248
http://www.elsevier.com

This academic journal publishes original research papers in the area of materials science. It specifically focuses on solar cells, photothermal devices, photochemical devices, and energy systems with innovative designs in light control and optical properties. The audience for this journal is primarily physicists, electrochemists, and energy researchers.

Wind Energy
John Wiley & Sons Ltd.
Corporate Headquarters
111 River Street
Hoboken, NJ 07030-5774

ISSN: 1095-4244

This journal publishes academic and industrial research, reports, and reviews concerning developments in wind power and technology. It also publishes research in environmental, economic, and sociopolitical aspects of wind energy technology. The journal is aimed toward professionals and academics in the field of wind energy.

Films and Videorecordings

Alternative Power

Produced by ITV in conjunction with the U.S. Department of Energy's Offices of Advanced Automotive Technologies, Energy Efficiency and Renewable Energy, and Fossil Energy and Enron Wind Corp. Published by Films for the Humanities & Sciences, 2000.

Produced and written by K. Brown.
    VHS, 82 minutes

This educational film is about the transformation to renewable energy resources. It examines the emerging markets of wind turbines, solar cells, hybrids, and coal gasification technologies within the context of increasing global energy demand and global warming concerns. This is a good resource for those who want to lean more about renewable technologies and their role in changing the energy infrastructure in society.

Closing the Circle on Splitting the Atom: The History of the UMTRA Project.

Southwest Productions, 1995.
Directed by J. Cochran and B. Cox.
    VHS, 29 minutes

This video describes the development of the U.S. Department of Energy's Uranium Mill Tailings Remediation Action (UMTRA) cleanup projects. It describes the nuclear age and the problem of radioactive mill tailings left over from uranium mining, the health effects of radon and radioactive materials, the implementation of the U.S. government's UMTRA program, and ongoing cleanup activities.

Coal, Blood, and Iron

BBC TV production in association with the Arts & Entertainment Network, Coronet/MTI Films & Video, the Seven Network. Published by Coronet Film & Video, 1991.

Produced by M. Hughes-Games.
  VHS, 55 minutes

This film offers a historical overview of the discovery and development of coal resources in Europe at the dawn of the industrial revolution. It examines the social and economic implications of coal and how the industrial revolution changed life in Europe leading up to World War I.

Electric Nation

Great Projects Film Company, Inc., in association with South Carolina ETV and the National Academy of Engineering. Distributed by PBS and published by Great Projects Film Company, Inc., 2002.

Produced and written by D. A. Miller.
  VHS, 60 minutes

This film examines the historical roots of the electricity industry in the United States. It tells the story of Thomas Edison, the invention of the incandescent lightbulb, the rise of electric utility companies, and the construction of the Tennessee Valley Authority.

Hostages to Oil

BBC TV production in association with the Arts & Entertainment Network, Coronet/MTI Films & Video, the Seven Network. Published by Coronet Film & Video, 1991.

Produced by M. Andrews.
  VHS, 55 minutes

This film provides an overview of the energy crisis in Europe. It examines the policies of energy independence and diversification, the development of North Sea oil, the relationship between Europe and the Middle East, and the problems with oil dependency.

Power and Frontiers

BBC TV production in association with the Arts & Entertainment Network, Coronet/MTI Film & Video, the Seven Network. Published by Coronet Film & Video, 1991.

Produced by M. Andrews.
VHS, 55 minutes

This film examines the role of oil and coal in the political, social, and economic developments that occurred in Europe during the twentieth century. It describes the role of oil in World War II and the importance of coal for sustaining daily life in the early twentieth century.

The Prize: The Epic Quest for Oil, Money and Power

InVision production for Majestic Films International in association with BBC Television, MICO, and WGBH Boston. Published by Public Media Video, 1993.

Produced and directed by W. Cran and S. Tepper.
VHS, 8 hours, 4 videocassettes

This film series tells the story of the rise and dominance of the oil industry throughout the world. It provides a detailed historical overview of oil companies, their political and economic influence, and the impact of oil in the development of modern society.

Databases and Internet Sites

Bioenergy Feedstock Information Network (BFIN)
http://bioenergy.ornl.gov/

This Web site provides information on biomass energy resources and supply systems. It offers an interactive database where users can select the type of biomass resource (e.g., herbaceous crops, forestry residue, etc.) and the process stage (e.g., harvesting) and obtain links to reports, databases, fact sheets, and presentations related to that resource. The Web site also provides an overview of the various forms of biomass energy, the economic and environmental considerations of biomass, and research and development initiatives.

Centre for Analysis and Dissemination of Demonstrated Energy Technologies (CADDET)
http://www.caddet.org/

This Web site provides a searchable database of information about energy efficiency and renewable energy technologies. It provides information about efficiency technologies used in industry, building transport, utilities, and agriculture; knowledge about energy-saving measures in all sectors; and suppliers and manufacturers of equipment and technology.

Database of State Incentives for Renewables and Efficiency (DSIRE)
http://www.dsireusa.org/

This Web-based database lists state and local incentives offered for the use of renewable technology and energy efficiency measures for each state. Information given includes financial incentives for renewable and energy efficiency measures and rules, regulations, and policies that apply to renewable programs in each state. Information for federal programs is also available.

Global Energy Law Portal
Oil, Gas and Energy Law Intelligence (OGEL)
http://www.gasandoil.com/ogel/

This Web site provides up-to-date information about regulations, legal instruments, and standards imposed and recommended by various jurisdictions (e.g., countries, international organizations, regional governments, etc.). It offers a database of legal and regulatory materials and current developments in the oil and gas industry.

Integrated Nuclear Fuel Cycle Information Systems
International Atomic Energy Agency (IAEA)
   http://www.nfcis.iaea.org/Default.asp

This database contains current information and statistics on the nuclear fuel cycle in countries that use nuclear energy. It provides information about all steps in the nuclear fuel cycle, from uranium reserves estimates to fuel processing to waste handling and storage. It is intended for member states, the IAEA, and the citizens who are interested in tracking the nuclear fuel cycle.

National Coal Resource Data System (NCRDS)
U.S. Geological Survey
   http://energy.er.usgs.gov/products/databases/
   USCoal/index.htm

This database contains coal resource data for coal regions in the United States. It provides information about the location, bed thickness, moisture, sulfur content, heat value, and general chemistry of coal deposits found in the United States. It also gives updated information about coal resource and reserve estimates.

National Renewable Energy Laboratory (NREL) Publications Database
   http://www.nrel.gov/publications/

The *NREL Publications Database* contains publications and other materials developed and written by NREL. It provides electronic access to technical reports, journal articles, conference papers, patents, presentations, and other print and nonprint resources on research efforts in renewable energy.

Oil Voice
   http://www.oilvoice.com/m/default.asp

The Oil Voice Web site offers a worldwide view of the oil and gas industries. It provides company profiles, news releases, financial news and statistics, updated production and reserves numbers, and an overview of the important people and places in the oil industry.

Energy Problems and Solutions: Economics and the Environment

Books

Boyle, G., B. Everett, and J. Ramage, eds. 2003. *Energy Systems and Sustainability*. Oxford, England: Oxford University Press. 619 pp.

This book provides an overview of energy systems and their social, political, economic, and environmental impact. It examines the various energy resources used throughout most of the world, provides a basic overview of how energy systems work, explains the economics involved in energy provision and distribution, and assesses the use of these resources within the context of sustainability. This is a good reference for students who require a broad overview of current energy dynamics.

Claussen, E., ed. 2001. *Climate Change: Science, Strategies, and Solutions*. Pew Center on Global Change. Leiden, Netherlands: Brill. 399 pp.

This book is a useful overview on the issue of global warming. It provides a summary of the science, economics, global strategies, and solutions associated with climate change. It compiles information provided by top researchers around the world and is an ideal reference for students and scholars who need to understand all aspects of global climate change.

Convery, F. J. 1998. *A Guide to Policies for Energy Conservation: The European Experience*. Cheltenham, England: Edward Elgar Publishing. 169 pp.

This book analyzes energy conservation initiatives and policies in Europe. It examines case studies of government investment and subsidies in energy conservation measures, the European Union's experience with combined heat and power, demand-side management, and the role of institutions in energy conservation. This book is a useful reference for students who wish to understand the impact of energy conservation policies.

Doyle, J. 2000. *Taken for a Ride: Detroit's Big Three and the Politics of Pollution.* New York: Four Walls Eight Windows. 560 pp.

This book examines the influence of large automobile manufacturing companies on air quality policies, fuel efficiency, and emission standards imposed by the federal government. It describes the lobbying tactics used by General Motors, Ford, and Chrysler (the "Big Three" of the auto industry) to delay the implementation of pollution enforcement and the ratification of the Kyoto Protocol. This book is a good resource for students who are interested in corporate influence on environmental regulations and provides an interesting investigative case study into the links between business and government.

Elliot, D. 1997. *Energy, Society and Environment: Technology for a Sustainable Future.* London, England: Routledge. 252 pp.

This book examines the environmental impacts of energy use in society and potential solutions. It reviews major environmental problems (e.g., air pollution, acid deposition, etc.), examines the technological "fixes" that have been introduced to deal with these problems, describes sustainable alternatives to fossil fuels and the issues of implementing sustainable programs, and outlines the necessary steps for addressing energy problems by building sustainable societies. This book is a good reference for those who wish to understand the link between energy technologies and sustainable progress.

Ewing, R. A., and D. Pratt. 2005. *Got Sun? Go Solar: Get Free Renewable Energy to Power Your Grid-Tied Home.* Masonville, CO: PixyJack Press. 159 pp.

This book is a practical how-to guide to the installation of renewable energy systems in typical residential dwellings. It covers equipment requirements, legal issues, incentives and rebates, and permit requirements. It serves as a useful guide for people wanting to use alternative energy sources in their grid or nongrid-connected homes.

Gellar, H. S. 2003. *Energy Revolution: Policies for a Sustainable Future.* Washington, DC: Island Press. 289 pp.

This book examines the technological, policy, and economic considerations required for building sustainable energy systems. It examines barriers that prevent the dissemination of energy-efficient technologies, discusses policy incentives and market reforms for promoting sustainable uses of energy, and analyzes successful, sustainable energy measures in developing and industrialized countries using Brazil and the United States as examples. This book is a useful resource for those who are interested in effective policy measures and market reforms for promoting alternative technologies.

Grubb, M., C. Vrolijk, and D. Brack. 1999. *The Kyoto Protocol: A Guide and Assessment.* **London, England: Royal Institute of International Affairs, Energy and Environmental Programme. 342 pp.**

This book is an assessment of the Kyoto Protocol. It provides an overview of the scientific, political, and legal foundations of the document, profiles of the players involved in the Kyoto negotiations, a description of the various commitments outlined for each of the players, an overview of mechanisms such as joint implementation and the clean development mechanism, and a projection of the challenges faced in implementing the Protocol. This book is a useful resource for anyone who wishes to understand the Kyoto Protocol.

Kaya, Y., and K. Yokobori, eds. 1997. *Environment, Energy, and Economy: Strategies for Sustainability.* **Tokyo, Japan: United Nations University Press. 381 pp.**

This book examines issues associated with development in the context of energy use, economics, and environmental problems. The contributors focus largely on climate change and potential impacts to energy markets in both developing and industrialized countries. Other topics look at the energy-economic interface, the social barriers that impede sustainable development, increasing energy consumption in developing countries, and technical developments in the area of decarbonization policies, leapfrogging strategies, and technology transfer. This book is an excellent reference for students who need an integrated and globalized approach to understanding energy in the context of economics and the environment.

Khagram, S. 2004. *Dams and Development: Transnational Struggles for Water and Power.* Ithaca, NY: Cornell University Press. 270 pp.

This book examines the issues associated with the construction of large dams for the purposes of hydropower and development. Using India's Narmada Projects as a case study, the author extrapolates his analysis of antidam movements to similar dam projects in other developing countries. Since water projects provide a substantial amount of electricity in the developing world and are likely to be pursued in the future, this book provides a timely overview of the human rights and environmental concerns associated with building large dams.

Kursunoglu, B. N., S. L. Mintz, and A. Perlmutter, eds. 2001. *Global Warming and Energy Policy.* New York: Kluwer Academic/Plenum Publishers. 220 pp.

This book examines the issue of global warming within the context of energy policy. Contributors focus mainly on the issue of nuclear energy, its potential development in response to climate change concerns, the problems associated with heavier reliance on nuclear resources, and how trends of market restructuring in the electric utilities industry could impact nuclear power plant performance. This book is a good reference for those who want to understand the role of nuclear energy in the face of global climate change concerns.

Leggett, J. K. 2005. *The Empty Tank: Oil, Gas, Hot Air, and the Coming Global Financial Catastrophe.* New York: Random House. 236 pp.

This book examines the issue of a global energy crisis resulting from depleting sources of oil and natural gas. It describes the potential impact to financial markets and governments worldwide, explains how energy companies have avoided the issue of an energy crisis, and supports the development and widespread use of innovative energy technology to avoid a financial crash. This book is a good reference for those who are interested in the peak oil debate.

Lemco, J., ed. 1992. *The Canada–United States Relationship: The Politics of Energy and Environmental Coordination.* New York: Praeger. 222 pp.

This book examines the cross-border linkages between the United States and Canada in the areas of energy and environmental policy. It provides a comparative analysis of the policy approaches in each country, the transboundary conflict between the two regimes, the roles of business and government in regulating energy industries, and how environmental groups have impacted the energy sector in each country. This book is a useful resource for understanding the link between energy and environmental policy and how this link impacts relations between Canada and the United States, two important players that have shaped the energy dynamics of North America.

Morgenstern, R. D., and P. R. Portney, eds. 2004. *New Approaches on Energy and the Environment: Policy Advice for the President.* **Washington, DC: Resources for the Future. 154 pp.**

This book offers a number of innovative energy policy approaches to environmental problems. It touches on measures that impose carbon and gasoline taxes, support fuel efficiency initiatives, clean up emissions from power plants, refine air quality standards, promote green energy, and decrease U.S. reliance on oil. This book offers an excellent reference for students who want to learn about effective and equitable policy initiatives for environmental issues associated with energy use in society.

Pinderhughes, R. 2004. *Alternative Urban Futures: Planning for Sustainable Development in Cities throughout the World.* **Lanham, MD: Rowman & Littlefield. 272 pp.**

This book discusses policy approaches and technologies that can be used in the planning and development of urban areas. It focuses on water and waste management, energy production and use, and transportation and food systems, providing an overview of innovative technologies in each area. It examines ways in which city design and management can provide a high quality of life that is environmentally sustainable. This book is a good reference for those who need a practical approach to energy systems management in urban societies.

Rao, P. K., ed. 2000. *The Economics of Global Climate Change.* **Armonk, NY: M. E. Sharpe. 199 pp.**

This book provides a foundation for understanding economic aspects associated with global climate change. It discusses potential global economic consequences of climate change, proposed economic measures for reducing greenhouse gas emissions, and international institutional mechanisms such as joint implementation and clean development mechanisms, and then summarizes the use of these measures in the Kyoto Protocol. This book is a useful reference for students, scholars, and policymakers who wish to understand the link between economics and global climate change.

Roberts, P. 2004. *The End of Oil: On the Edge of a Perilous New World*. Boston, MA: Houghton Mifflin. 399 pp.

This book examines the issue of global peak oil production and the social and economic consequences that could result from this problem. It highlights important themes such as energy literacy, or the need for heightened awareness of energy issues, and the need for increased energy efficiency. It presents potential scenarios that could result from higher oil prices and continued depletion. This book is a good resource for students who need to understand the issue of peak oil and the arguments for energy reform in society.

Rodgers, W. M., Jr. 2000. *Third Millennium Capitalism: Convergence of Economic, Energy and Environmental Forces*. Westport, CT: Quorum. 278 pp.

This book promotes the institution of democratic capitalism for meeting global economic and energy needs in the twenty-first century. It describes the structure of global corporations and markets, examines world energy demand, supply, and provision within the current economic structure, and links environmental issues to future forecasts of economic growth and development. This book is a useful resource for students who wish to understand global energy market dynamics and how they relate to economic development and environmental stewardship.

Schneider, S. H., A. Rosencranz, and J. O. Niles, eds. 2002. *Climate Change Policy: A Survey*. Washington, DC: Island Press. 563 pp.

This book provides a useful overview of global climate change. It examines the issue from a broad array of perspectives, including climate science; regional impacts; international approaches for reducing greenhouse gas emissions; business and economic aspects; how climate change may affect agriculture, human populations, and tropical forests; the role of renewable technologies and carbon sequestration policies; and how the issue can be addressed in an equitable manner. This book is an excellent reference for those who want to understand the issue of climate change by studying different perspectives.

Smith, E. R. A. N. 2002. *Energy, the Environment, and Public Opinion.* **Lanham: MD: Rowman & Littlefield. 264 pp.**

This book examines how Americans have viewed energy issues since World War II, through periods of energy abundance and scarcity. It discusses energy supplies, the media portrayal of energy resources, the impact of environmental disasters—such as the oil spill off the coast of Santa Barbara, California—on public opinion, and how attitudes about energy and the environment relate to the various theories on public opinion and policy. This book is a useful reference for students interested in studying the interaction between public opinion and environmental policy.

Smith, Z. A. 2004. *The Environmental Policy Paradox.* **4th ed. Upper Saddle River, NJ: Prentice Hall. 295 pp.**

This book provides a comprehensive overview of environmental policy in the United States. It explains the institutional and legal settings for policy formation, examines the environmental problems in all types of media (i.e., air, land, water), and devotes a chapter to energy policy. This book is a useful resource for students and policymakers who wish to understand how environmental policy is made and implemented in the United States.

Swartz, M., and S. Watkins. 2003. *Power Failure: The Inside Story of the Collapse of Enron.* **New York: Doubleday. 386 pp.**

This book provides a historical account of the rise and fall of the Enron Corporation. It explains how Enron became one of the most powerful providers of energy services in the United States, how it manipulated energy markets to benefit shareholders, and what caused its ultimate collapse. This book is a useful resource for students studying the issue of electricity regulation in the United States and the problems that arise when states deregulate their electricity markets.

Sweet, W. 2006. *Kicking the Carbon Habit: Global Warming and the Case for Renewable and Nuclear Energy.* New York: Columbia University Press. 256 pp.

This book critically examines the United States' role in global climate change and its approach to reducing greenhouse gas emissions. It discusses the adverse effects of global warming, summarizes recent research from leading scientists, and advocates for the United States to reduce its reliance on coal resources for electricity and instead promote the development of renewable and nuclear technologies. This book is a useful, readable update of global warming science and offers an argument for reducing the use of coal-based technologies.

Periodicals, Journals, and Newsletters

Annual Review of Environment and Resources
Annual Reviews Inc.
4139 El Camino Way
P.O. Box 10139
Palo Alto, CA 94303-0139

ISSN: 1543-5938

This academic journal publishes research, reviews, and reports on the impact of resource use on the environment. Topics include climate change, human impact, environmental and energy management, and sustainable development.

Home Energy Magazine
Energy Auditor and Retrofitter
2124 Kittredge Street #95
Berkeley, CA 94704

ISSN: 0896–9442
http://www.homeenergy.org/hewebsite/

This monthly magazine provides information for people interested in designing and building energy systems for residential dwellings. It covers issues such as efficiency, performance, and comfort.

International Journal of Sustainable Development and World Ecology
Sapiens Publishing
Duncow
Kirkmahoe
Dumfriesshire DG1 1TA, United Kingdom
   ISSN: 1350-4509

This scholarly interdisciplinary journal publishes research on sustainable development in the disciplines of biology, environmental sciences, sociology, political science, economics, and law. It seeks to provide in-depth coverage of the meaning of sustainable development in diverse fields.

International Journal of Sustainable Energy
Taylor and Francis Ltd.
4 Park Square
Milton Park, Abingdon, Oxfordshire
OX14 4RN, United Kingdom
   ISSN: 1478-6451

This academic journal is targeted toward scientists and engineers who are studying technological innovations in the area of sustainable energy. It describes research conducted in both developing and industrialized nations in all areas of renewable energy.

International Journal of Technology Management and Sustainable Development
Intellect Ltd.
The Mill, Parnall Road
Fishponds, Bristol BS16 3JG
United Kingdom
   ISSN: 1474-2748

This scholarly journal publishes up-to-date research on the interaction between technology and development. It focuses mainly on technology in developing countries, environmental sustainability, research and development strategies, knowledge and capacity building, and technology transfer.

Oil and Chemical Pollution
Elsevier, Inc.
30 Corporate Drive, 4th floor
Burlington, MA 01803
ISSN: 0143-7227

This journal publishes research and review papers in the areas of pollution remediation and impact of oil and chemical spills. It provides analysis of the environmental and ecological impacts of oil spills, oil spill dispersion dynamics, effectiveness of new treatments and technologies, and surveys of the effects of large oil spills.

Oil Spill Intelligence Report
Aspen Publishers
76 Ninth Avenue, 7th Floor
New York, NY 10011
ISSN: 0195-3524

This weekly journal provides reports on oil spills across the world; information about contingency planning, scientific research, new technology in the areas of cleanup, control, and prevention; and news of international efforts to address the issue of oil spills. It targets a wide audience of industry, academic professionals, and public officials involved in oil spill remediation.

Resource and Energy Economics
Elsevier, Inc.
30 Corporate Drive, 4th floor
Burlington, MA 01803
ISSN: 0928-7655

This academic journal publishes scholarly papers that analyze economic aspects of natural resource use and management, energy consumption, and environmental and energy policies in developing and industrialized countries. It is an interdisciplinary journal for people studying links between energy, the environment, and economics.

Films and Videorecordings

Building Sustainability with the Natural Step
Northcutt Production, 1999.

Produced and directed by P. Northcutt.
VHS, 22 minutes

This film examines the concept of sustainability and the importance of preserving resources for the needs of future generations. It shows how the concept of sustainable living has been incorporated into facilities at the University of Texas in Houston.

Building with Awareness: The Construction of a Hybrid Home
A Syncronos Design Production, 2005.

Written, produced, and directed by T. Owens.
DVD, 162 minutes

This film is about green building technology for domestic dwellings. It discusses straw bale, adobe, and cob building materials; describes construction techniques and tips; and goes over things that need to be considered in the planning, design, and construction of green homes, including passive heating and cooling systems, the use of photovoltaic technology, and rainwater cisterns.

Can Polar Bears Tread Water? The Changing Climate

A Central Television/TVE production, produced in association with the Television Trust for the Environment and the Better World Society, 1989. Published by MTI Film & Video.

Produced and directed by L. Moore.
VHS, 53 minutes

This film examines the issue of global warming. It describes the greenhouse effect, discusses the causes and consequences of global warming, including the impact to coastal areas and developing countries, and reviews international policy initiatives, such as the Montreal Protocol, that have been initiated to deal with the issue of global warming.

Crisis in the Atmosphere

WQED Pittsburgh in association with the National Academy of Sciences, 1989. Published by Intellimation.

Written and produced by L. Friedberg.
VHS, 60 minutes

This film examines the scientific evidence for global warming. It provides an overview of the issue and the potential consequences to the quality of life in industrialized and developing countries.

Dennis Weaver's Earthship

Robert Weaver Enterprises production in association with Gerry Productions, 1990. Published by Survival Habitat.

Produced and written by R. W. and M. T. Scarpaci; directed by P. Scarpaci.
VHS, 29 minutes

This film describes the construction of "Earthship," a home made out of recycled tires, sand, mud, and soda cans. It describes how the house produces and distributes energy in an efficient way from solar technologies and examines how passive solar elements are used for heating and lighting.

The Energy Bank

Umbrella Films and Rampion Visual Productions. Published by Bullfrog Films, 1991.

Produced and directed by A. C. Grossman and R. F. Cole.
VHS, 38 minutes

This film discusses the need for enhancing energy efficiency in our society. It examines efficiency in electricity generation, consumption, and transportation, explains the energy gains that can be made through energy efficiency measures, and describes the political and institutional barriers to implementing efficiency measures.

Enron: The Smartest Guys in the Room

Magnolia Pictures, HDNet Films Presentation and Jigsaw Productions. Published by Magnolia Home Entertainment, 2005.

Produced by A. Gibney, J. Kliot, and S. Motamed; written and directed by A. Gibney.
DVD, 110 minutes

This documentary film is about the collapse of one of America's largest energy trading companies. It examines Enron's rise to power, its accumulation of energy utilities, and the corrupt financial activities and manipulations in the energy trading industry.

The Four Corners: A National Sacrifice Area?

Bullfrog Films, in association with the Graduate School of Journalism, University of California, Berkeley. Published by Bullfrog Films, 1996.

Produced by C. McLeod, G. Switkes, and R. Hayes, written and directed by C. McLeod.
VHS, 58 minutes

This documentary examines the impacts of coal and uranium mining on the Navajo and Hopi tribes in the southwestern United States. It describes the environmental degradation from coal mining, the health problems associated with uranium mining, and the environmental and cultural impacts of resource extraction.

An Inconvenient Truth
Paramount Classics and Participant Productions, 2006

Directed by D. Guggenheim and produced by L. David, L. Bender, and S. Z. Burns.
DVD, 100 minutes

This film features the campaign of former vice president Al Gore to educate the public about global climate change and confront issues of global warming. It describes the science of global warming in a tangible way and encourages viewers to participate in actions that mitigate the environmental consequences of energy use.

Oil on Ice

Dale Djerassi/Bo Boudart Production in association with Lobitos Creek Ranch. Published by Oil On Ice Partners, 2004. Distributed by Bullfrog Films.

Produced and directed by D. Djerassi and B. Boudart; written by S. Most.
DVD, 90 minutes

This documentary concerns issues associated with drilling for oil in the U.S. Artic National Wildlife Refuge (ANWR). It examines the livelihood of the Gwich'in Indians, the potential impact that drilling may have on migratory bird and caribou populations, and the debate among environmentalists and oil companies. This film is a good resource for people who want to understand the ANWR drilling issues and see stunning visual footage of the refuge.

Databases and Internet Resources

Alternative Fuels Data Center (AFDC)
http://www.eere.energy.gov/afdc/

The AFDC provides up-to-date information on alternative fuels and vehicles used in the transportation sector. It contains information on biodiesel, electric, ethanol, hydrogen, and natural gas fuel sources, offers an updated listing on available alternative fuel stations, and lists more than 3,000 documents relating to alternative fuels.

Best Practices Database for Improving the Living Environment
http://www.bestpractices.org/

This searchable database contains solutions for public and private entities working to improve governance, eliminate poverty, protect the environment, and enhance economic development. The database includes policy tools, networking and technical cooperation opportunities, and analysis of current trends in sustainable development.

European DataBank Sustainable Development
http://www.sd-eudb.net/

This database is for institutions, organizations, and associations involved in planning and implementing sustainable development measures in their societies.

Eurostat, Sustainable Development Indicators
European Union
http://europa.eu.int/comm/eurostat/

This database provides information and statistics on a number of indicators of sustainable development for policymakers, the general public, and academic communities. Themes include economic development, production and consumption patterns, management of natural resources, and transport.

National Sustainable Agricultural Information Service (ATTRA) and National Center for Appropriate Technology (NCAT)
http://attra.ncat.org/

This Web site offers information on sustainable energy practices and energy-efficient agricultural practices. Topics include sustainable farming technologies, practices, and programs for farmers, ranchers, researchers, and educators.

United Nations Common Database (UNCDB)
http://unstats.un.org

This database provides a large amount of statistical information in the areas of global finance, national accounts, food and agriculture production, industrial commodities, and world development indicators for countries and regions worldwide. It is a useful resource for people involved in global sustainable development research.

Glossary

Achnacarry Agreement ("As-Is") An agreement among the major oil companies to fix oil prices using the "Gulf Plus System" that priced oil as if it had been shipped from the Gulf of Mexico.

Acid rain *See* Atmospheric deposition.

Air pollution Gaseous and solid particles released in the process of fossil fuel combustion. Common air pollutants include nitrous oxide, sulfur dioxide, carbon monoxide, and particulate matter.

Anaerobic digestion A process that occurs during bacterial decomposition in anoxic environments of a biomass energy resource producing methane as a fuel.

Anthracite A type of coal characterized by high carbon content. This coal is mainly used for domestic heating purposes.

Atmospheric deposition Deposition of liquid (e.g., acid rain) and solid forms (lead, mercury, etc.) of air pollution onto land surfaces. The difference between deposition and precipitation is that deposition includes atmospheric fallout that is not in liquid form and can occur when it is not precipitating.

Atomic bomb An extremely powerful explosive device that causes a nuclear chain reaction when it is detonated.

Barrel of oil equivalent (boe) A unit of energy that specifies how much potential energy is contained in a barrel of oil. It is estimated to be 43.8 GJ of energy.

Big Inch The first large oil pipeline built to deliver crude oil from the southwestern United States to Pennsylvania.

Biofuels Fuels made from dried biomass (e.g., dung, wood, etc.) or from thermochemical processes (e.g., fermentation, anaerobic digestion, etc.) that transform biomass resources into fuels.

Biomass The living and/or biological material found on the earth's surface. It is often combusted or burned as an energy source.

Bituminous coal A ranking of coal that is characterized by lower carbon content. It is the primary type of coal used in electricity generation.

British thermal unit (Btu) A unit of energy that describes the quantity of heat needed to raise the temperature of one pound of water by one degree Fahrenheit.

Calorie A unit of energy that describes the amount of energy required to heat one gram of water one degree Celsius.

Catalytic cracking A petroleum-refining process that uses high temperatures and pressures to break large hydrocarbon molecules into smaller constituents.

Charcoal Carbon substance produced when wood undergoes pyrolysis. Historically, it was important for the smelting of iron ore.

Coal A chemically complex fossil fuel ranked into three major groups (bituminous, lignite, and anthracite) according to the amount of fixed carbon and volatile matter contained in its chemical structure.

Cogeneration A process for generating electricity from waste heat that is a by-product of electricity generation.

Coke Carbon fuel produced from the pyrolysis of coal. Historically, it was important for the manufacture of steel.

Combustion A process of releasing the chemical energy stored in fossil fuels and hydrocarbons by heating.

Commercial sector The energy sector that encompasses heating, cooling, and lighting of businesses, schools, hospitals, and churches.

Consumerism The trend of rapidly increasing demand and expanded distribution for a large number of goods and services resulting in an increased demand for energy.

Cord of wood 128 cubic feet of wood equaling a wood stack that is 4 feet × 4 feet × 8 feet.

Crude oil The thick, viscous petroleum compounds that are pumped from the ground and refined into gasoline and other fuels.

Deep shaft mining A method of extracting coal from the earth in areas where coal seams are located 100 feet or greater below the surface.

Deregulation The process of repealing government regulations on industry for the purpose of promoting economic growth and competition in markets.

Distillation A process that separates and collects hydrocarbon products in petroleum using their different boiling points.

Energy   The capacity to do work.

Energy crisis   Significant shortages in energy resources resulting in loss of energy services, spikes in energy prices, and overall decrease in energy use.

Energy conversion   The process of converting energy from one form to another (e.g., potential energy to mechanical energy, mechanical energy to electrical energy, etc.).

Energy dynamics   The interaction of society's energy systems with its economic and social structures.

Energy efficiency   The ratio of useful energy output to total energy input.

Energy flows   The path that energy follows as it is extracted, converted, delivered, and used in society.

Energy infrastructure   The physical structures that are built to deliver energy services to a society.

Energy intensity   A measure that describes energy use as a function of gross domestic product (GDP). It is often used to describe energy trends in a particular country.

Energy sector   A categorization of end uses of energy that group into four divisions: residential, commercial, industrial, and transportation.

Energy security   Condition which is met when energy resources are available, affordable, and reliable to energy consumers in a society.

Energy subsidies   Payments or rewards granted by governments to energy companies for the purpose of minimizing the cost of energy production for public provision.

Energy transition   A shift of the primary energy resource used in large societies (e.g., coal to petroleum).

Energy units   A particular quantity of energy, power, or resources associated with energy use that is accepted as a standard for measurement or trade (e.g., joule is the standard unit of energy).

Engineering   A field that utilizes physical laws to design systems for harnessing and distributing energy to society.

Federal lands   Land areas in the United States that are publicly owned and managed by federal government agencies.

Fermentation   A process that occurs during the breakdown of biomass resources involving aerobic microorganisms to produce biofuels (e.g., ethanol).

Force   The product of mass and acceleration (mass × acceleration).

Fossil fuels (energies) Formed from geologically compressed layers of organic matter and composed of mostly light elements such as carbon, nitrogen, and oxygen, they are used as an energy resource through a process called combustion. Petroleum, coal, and natural gas are the main fossil energies used by society.

Fuel cycle The complete process of extraction, transportation, processing, and consumption of a fuel type (e.g., coal, nuclear fuel rods, etc.).

Gas flaring A highly polluting practice that removes unwanted gas from crude oil.

Gasification The process by which a gaseous fuel is produced from a solid using steam.

Generator A machine that produces electrical energy from mechanical energy by passing a coil of conductive wire past the positive and negative poles of a magnet.

Geothermal energy A primary energy source derived from hot subsurface environments and used to create steam for electricity generation.

Global energy market The means by which energy resources are traded among countries and within entities that supply, produce, and distribute energy.

Global warming Warming of the Earth's surface temperature by several degrees. Warming occurs as solar radiation is trapped by gases in the Earth's atmosphere. Warming is a natural phenomenon but can be enhanced by human activity.

Globalization The integration of regional markets across national boundaries.

Greenhouse gases Gases emitted during fossil fuel combustion that absorb infrared radiation from the Sun and hence do not allow solar radiation to escape the Earth's atmosphere.

Gross domestic product (GDP) A measure of the goods and services produced annually in a particular country.

Hydrocarbons Molecules composed of carbon and hydrogen chains. They form the chemical basis of all fossil energies.

Induction The process by which electrical current is generated in a charged circuit from an adjacent charged circuit by proximity and grounding.

Industrial revolution A period in history that marks the rise in manufacturing and industry.

Industrial sector The energy sector that describes energy use for purposes of manufacturing, textiles, paper industries, metallurgy, chemical industries, and oil refining.

Isotopes Atoms of the same element that contain different numbers of neutrons in their nuclei.

Joule The international standard unit (SI) for energy.

Hydroelectric dam A large structure built across a river for the purpose of capturing the kinetic energy of falling water to generate electricity.

Hydroelectricity Electricity generated from the kinetic energy of falling water.

Kerosene A product distilled from crude petroleum. Originally used as an illuminant, it was the first attractive commodity to be separated from petroleum.

Kilowatt-hour A common unit for electricity rates that expresses the amount of electricity consumed in one hour.

Land reclamation The process of restoring lands decimated by coal mining activities to their original ecological integrity.

Land subsidence An occurrence where land sinks down and sometimes collapses into abandoned mine shafts.

Law of Thermodynamics (1) Conservation of energy; energy cannot be created or destroyed. (2) As energy is converted from one form to another, it becomes less useful for doing work.

Liquefied natural gas (LNG) Natural gas cooled to −259°F for storage and transportation. It is regasified at its destination before use.

Market liberalization Occurs in globalized markets when trade borders for goods and services are relaxed to promote the interests of transnational corporations.

Market privatization Occurs when nationalized (state-owned) industries, such as oil industries, are sold to transnational corporations.

Marshall Plan A policy measure enacted by the United States with the intention of providing economic and energy aid to Europe after World War II.

Monopoly Condition described as a market failure that occurs in a market system where only one or a few providers of a good or a service exist, thereby limiting consumer choice.

National Ambient Air Quality Standards (NAAQS) Limits established by the U.S. government for six primary air pollutants: SO_2, NO_x, ozone, carbon monoxide, lead, and particulate matter.

Natural gas A fossil fuel primarily made up of methane, a compound composed of molecules containing one carbon and four hydrogen atoms.

Nonrenewable energy sources Energy sources that are depleted through overuse and replenished over long periods of geologic time. Examples include coal, oil, natural gas, and uranium.

Nuclear energy An energy source that is derived from nuclear fission, a process that utilizes neutrons to split uranium and plutonium atoms.

Offshore drilling Oil and natural gas drilling activities that occur along oceanic continental shelf regions.

Oil *See* Petroleum.

Oil embargo The imposition of sanctions by an oil-producing country that refuses to trade oil with the consuming country.

Petroleum (oil) Composed of hundreds of different hydrocarbons, this liquid resource is refined to produce useful consumer products and fuels (e.g., gasoline, kerosene, jet fuel, petroleum jelly, etc.).

Physics Science that explains processes and phenomena describing energy dynamics observed in the world.

Photovoltaics A class of compounds that convert light directly to electricity using solid-state crystalline materials.

Pipeline A common method for transporting fossil fuels that uses high pressures and mechanical pumps to push gas or liquid fuels through a network of pipes to destinations where they are to be converted into useful energy.

Power The rate at which energy is converted into electricity. It is measured in watts (joules/seconds).

Primary energy resource The fundamental energy source that provides a community with energy.

Pyrolysis The process of heating an energy source in the absence of air. It is used to make charcoal out of biomass resources and coke from coal resources.

Radiation The physical form of all free energy (e.g., light energy, microwave energy, ionizing energy, etc.).

Radioactive waste Unwanted radioactive by-products that result from nuclear reactions, such as those that take place in nuclear power plants.

Radioactivity A property of certain elements characterized by the spontaneous emission of energy in the form of rays or particles.

Refinery A processing plant that uses distillation and catalytic conversion to remove impurities and separate, collect, and purify useable products (e.g., gasoline, jet fuel, etc.) from petroleum.

Regulations Rules and guidelines imposed on industry by governments to ensure against the negative effects of market failures.

Renewable energy Natural energy flows, or sources, that are not significantly depleted with use and can be regenerated as they are depleted. Wind, solar energy, and water are examples of renewable energy sources.

Residential sector   The energy sector that describes energy used in homes for heating, cooling, lighting, electrical appliances, and cooking.

Semiconducting elements   Nonmetallic elements, such as silicon, that are able to conduct electricity.

Seven Sisters   The nickname given to the first seven global oil companies, so-named for their dominance in the global market and close alignment of interests.

Smelting   Procedure in steel-making that uses high temperatures to purify and strengthen iron ore.

Smog   A type of air pollution that forms when nitrogen oxide molecules react with ozone and water vapor in the atmosphere.

Solar energy   Energy that is contained in solar radiation.

Steel   A product made from the smelting of iron using coke as the smelting fuel. In this process, impurities are removed from the iron ore resulting in a structural material that is stronger than pig iron. Pig iron is the crude iron product that is retrieved from blast furnaces. It has a high carbon content and is more brittle than steel.

Strip (surface) mining   A method of coal extraction that involves stripping land away from the surface to reach coal seams located within 100 feet of the surface.

Sustainable development   An ideal way of life that embodies the concept that all people living on the planet have the same opportunities and resources available for enhancing their quality of life without compromising the ability of others to do the same. Practically, this concept involves increasing energy efficiencies, conserving energy, and developing renewable resources.

Thermochemical processing   A method of producing concentrated fuels from an energy resource (e.g., pyrolysis, gasification, etc.).

Transformer   Device that uses the principle of induction to transport electricity long distances from its generation source.

Transportation sector   The energy sector that encompasses energy used to transport people and goods from one place to another (e.g., trucks, railways, planes, etc.).

Uranium   The main radioactive fuel used in nuclear reactions.

Utility companies   Companies that generate and distribute energy services, such as electricity and natural gas.

Waterpower   Energetic power derived from the motion of falling water turning turbines.

Watt   A unit of energy where the rate of 1 watt is equal to 1 joule per second.

Wind energy   The use of wind to power turbines for the creation of electricity.

Work   The product of force and distance (force × distance).

Zircalloy   A metal alloy consisting of zirconium, tin, chromium, and nickel used as the casing in nuclear fuel rods because it is heat-resistant.

Index

About the Authors

Jaina Lorraine Moan received her B.S. in chemistry and biology and her M.A. in political science from Northern Arizona University. She is currently a research specialist for the Colorado Plateau Stable Isotope and Analytical Laboratories at Northern Arizona University. She has contributed to numerous research projects in the areas of chemistry, ecology, forestry, and environmental policy. She is avidly interested in interdisciplinary research that links aspects of environmental science, economics, law, and policy. This is her first book.

Zachary A. Smith received his B.A. from California State University, Fullerton, and his M.A. and Ph.D. from the University of California, Santa Barbara. He has taught at Northern Arizona University, the University of Hawaii, Ohio University, and the University of California, Santa Barbara, and served as the Wayne Aspinall visiting professor of political science, public affairs, and history at Mesa State College. A consultant both nationally and internationally on environmental matters, he is the author or editor of more than twenty books and many articles on natural resources and environmental topics. He currently teaches environmental and natural resources policy and administration in the public policy Ph.D. program in the Political Science Department at Northern Arizona University in Flagstaff. He encourages students interested in pursuing graduate studies in environmental policy to contact him.

DATE DUE

Demco, Inc. 38-293